ADVANCES IN CHEMICAL PHYSICS

VOLUME LXXVIII

Advances in
CHEMICAL PHYSICS

EDITED BY

I. PRIGOGINE

University of Brussels
Brussels, Belgium
and
University of Texas
Austin, Texas

AND

STUART A. RICE

Department of Chemistry
and
The James Franck Institute
The University of Chicago
Chicago, Illinois

VOLUME LXXVIII

AN INTERSCIENCE® PUBLICATION
John Wiley & Sons, Inc.
NEW YORK / CHICHESTER / BRISBANE / TORONTO / SINGAPORE

An Interscience® Publication

Library of Congress Catalog Number: 58-9935

ISBN 0-471-52666-5

Printed in the United States of America

10 9 8 7 6 5 4 3 2 1

CONTRIBUTORS

THOMAS L. BECK, Department of Chemistry, University of Cincinnati, Cincinnati, Ohio

L. BLUM, Department of Physics, University of Puerto Rico, Rio Piedras, Puerto Rico

J. D. DOLL, Department of Chemistry, Brown University, Providence, Rhode Island

DAVID L. FREEMAN, Department of Chemistry, University of Rhode Island, Kingston, Rhode Island

BONGSOO KIM, The James Franck Institute and Department of Physics, The University of Chicago, Chicago, Illinois

ANDRZEJ KOLINSKI, Department of Chemistry, University of Warsaw, Warsaw, Poland

GENE F. MAZENKO, The James Franck Institute and Department of Physics, The University of Chicago, Chicago, Illinois

JEFFREY SKOLNICK, Molecular Biology Department, Research Institute of Scripps Clinic, La Jolla, California

INTRODUCTION

Few of us can any longer keep up with the flood of scientific literature, even in specialized subfields. Any attempt to do more and be broadly educated with respect to a large domain of science has the appearance of tilting at windmills. Yet the synthesis of ideas drawn from different subjects into new, powerful, general concepts is as valuable as ever, and the desire to remain educated persists in all scientists. This series, *Advances in Chemical Physics*, is devoted to helping the reader obtain general information about a wide variety of topics in chemical physics, which field we interpret very broadly. Our intent is to have experts present comprehensive analyses of subjects of interest and to encourage the expression of individual points of view. We hope that this approach to the presentation of an overview of a subject will both stimulate new research and serve as a personalized learning text for beginners in a field.

ILYA PRIGOGINE
STUART A. RICE

CONTENTS

ADVANCES IN CHEMICAL PHYSICS

VOLUME LXXVIII

INTERNAL EXCITATIONS IN LIQUIDS

RICHARD M. STRATT*

Department of Chemistry, Brown University, Providence, RI

CONTENTS

I. INTRODUCTION

When the theory of liquids was first being constructed, it seemed trouble enough to deal with intermolecular correlations without having to worry about the additional complications created by intramolecular dynamics. A liquid composed of even completely featureless atoms would still lack the comforting periodicity displayed by crystalline solids, without having the advantage of the extreme level of disorder displayed by nonideal gases. The

1

first step, therefore, was to find a set of theoretical tools suitable for coping with this intermediate kind of order.

To a certain extent this goal has been achieved. The *structure* and *thermodynamics* of simple atomic and molecular liquids are now fairly well understood, in the sense that pair correlation functions (and hence neutron-scattering and equation-of-state results) can be predicted quite accurately from integral equations and from simulations, given a modest amount of information about the intermolecular potentials. Moreover, there is a simple physical picture underlying most of this success which attributes the structure largely to the geometry implicit in packing hard objects, such as spheres or dumbbells, in three dimensions—corrected, if necessary, by the energetics of interacting charges, dipoles, or higher multipoles. There is also a well-established diagrammatic approach to analyzing these problems, allowing one to predict the success (or failure) of the various analytical methods.[1,2]

Structure at this level (and its immediate thermodynamic consequences), however, does not begin to exhaust what we would like to know about liquids. Molecules can have different conformations in liquids than they do when isolated. They can also have different vibrational, electronic, and magnetic resonance spectra. In the extreme examples, such as collision-induced spectra, liquids will absorb light at frequencies either forbidden to, or totally irrelevant to, the individual molecules. Liquids also have some fundamental attributes, such as their overall electronic behavior, which are intrinsically bulk properties, but which cannot be predicted from their radial distribution functions or equations of state. This category includes not only such detailed information as the band structure (as might be measured by photoelectron spectroscopy), but also what should presumably be the glaringly obvious fact of whether the liquid is an insulator, a semiconductor, or a metal.

The very existence of properties of this sort is due to the presence of internal degrees of freedom in atoms and molecules. The Hamiltonians we should consider in order to understand these properties, therefore, are ones that include not just the fluctuating center-of-mass coordinates and Euler angles necessary to describe where the atoms and molecules are, but the relevant fluctuating internal coordinates: bond lengths and angles for vibrations, dihedral angles for conformation, and even the positions of the electrons for the electronic structure.

It will not in general be easy to be this literal in our representation of these internal coordinates. As we shall discuss in more detail, internal degrees of freedom tend to be very different in character from overall translation. Intramolecular statistical mechanics usually involves a faster time scale, a higher energy scale, and a much more quantum mechanical character—all of which pose their own technical problems for both analytical and simulation work. Nonetheless, it is the interplay between this same set of features and the liquid structure that gives rise to the very phenomena we want to study. So, we shall want to develop theoretical techniques and models suitable for focusing on precisely these points.

This review is devoted both to summarizing this development and to bringing together what has been learned about the general trends in intramolecular behavior in liquids from a variety of simple models systems. The emphasis is on a few kinds of explicitly quantum mechanical problems. In particular, there is no discussion of either conformational change or vibrations. It should also be noted that we are leaving out a number of topics that are not normally thought of in these terms, but that could, nonetheless, be profitably classified under the broad rubric of internal degrees of freedom in liquids. For example, there is no coverage of molecular rotation, largely because the intimate relation between orientation and liquid structure makes the problem seem more one of an "external" degree of freedom. We also omit any mention of chemical reactions in liquids, though transformations in molecular identity are formally equivalent to changes in internal state. On the other hand, there are several instances in which it is profitable to introduce totally artificial internal degrees of freedom into otherwise featureless atomic liquids, some of which are detailed here.

II. HAMILTONIANS

If we limit ourselves to single-component atomic liquids whose atoms have internal degrees of freedom, the Hamiltonian we want to consider will almost always be of the form

$$\hat{H} = \hat{H}_0(q) + V_I(q, R) + V_B(R). \tag{2.1}$$

Here q and R represent the set of internal coordinates and the set of atomic positions, respectively, for the N atoms in our liquid,

$$q = \{q_j; j = 1, \dots, N\},$$
$$R = \{\vec{r}_j; j = 1, \dots, N\},$$

$\hat{H}_0$ is what we call the *gas-phase* Hamiltonian, what the Hamiltonian would be if the atoms were isolated from each other

$$\hat{H}_0(q) = \sum_j \hat{H}_{0j}(q_j), \tag{2.2}$$

and V_B is the *bath* potential energy, what the sum of the interactions between the particles would be if the particles were featureless. Typically the bath potential is describable as a sum of pair potentials, $u_0(r)$

$$V_B(R) = \sum_{j<k} u_0(r_{jk}), \tag{2.3}$$

often composed in part, or in whole, of strongly repulsive interactions.

Of course the internal coordinates will behave differently in the liquid than they do in the gas phase only to the extent that internal degrees of freedom are coupled to the translation of the atoms. Frequently this coupling will also be expressible as a sum of pair potentials

$$V_I(q, R) = \sum_{j<k} v(r_{jk}; q_j, q_k), \tag{2.4}$$

depending on both the interparticle separation and the internal coordinates of the interacting atoms. In a surprising number of examples these pair interactions end up being bilinear in the internal variables, in either a scalar,

$$v(r_{jk}; q_j, q_k) = q_j q_k v(r_{jk}), \tag{2.5}$$

or a tensor sense

$$v(\vec{r}_{jk}; \vec{q}_j, \vec{q}_k) = \vec{q}_j \cdot \mathbf{T}(\vec{r}_{jk}) \cdot \vec{q}_k, \tag{2.6}$$

though it would be easy to imagine more complicated cases.

To anticipate the introduction of quantum mechanics, we have used carets over the quantities that are operators (and not just functions of the coordinates). Why, then, is Eq. (2.1) the relevant form—why is it generally unnecessary to include the kinetic energy of the bath and, more generally, why is it that only the intramolecular term needs to be thought of as an operator? Both features are intimately connected with the role that quantum mechanics plays in our systems. As we discuss in the next section, internal dynamics is so much more quantum mechanical than the atomic translation for our examples that the *adiabatic approximation* of neglecting the fact that translational kinetic energy is an operator is virtually exact.[3,4] So, just as in classical statistical mechanics, the center-of-mass momenta can be integrated out of the problem. The observation that the coupling term V_I usually commutes with the bath potential, though, is somewhat more subtle, and we shall return to it when appropriate.[5]

The generalization of Eq. (2.1) to mixtures is straightforward, but it is worth mentioning the particularly important class of examples in which there is a single solute, with its own internal degree of freedom, surrounded by feature-less solvent molecules. These situations display none of the interesting effects created by mutually interacting internal coordinates, but they are the simplest places to look for solvent effects on spectra and on quantum dynamics. Indeed, most of what is known about internal excitations in liquids falls under this heading. We shall find ourselves returning to these cases from time to time.

Having set up this broad framework, a few specific examples are in order.

A. Continuous Internal Coordinates

There have been quite a number of theoretical studies of conformation[6,7] and vibration[4,8-13] in solution and even studies of the coupled conformational and electronic phenomena that occur in conjugated polymer solutions.[14,15] Keeping to our pledge in the Introduction, however, our examples will lean toward models for electronic structure. The most obvious electronic structure model one could think of would be that of a single *hydrogenic atom in fluid*,[16] for which

$$\hat{H}_0 = \frac{\hat{p}^2}{2m} - \frac{e^2}{q}, \tag{2.7}$$

with m the reduced mass and e the electric charge. This idea of treating the electron positions as genuine fluctuating variables (along with its generalizations to more complicated atoms and molecules) has been exploited frequently in the context of continuum models for the liquid. Typically, the effect of the solvent is represented by putting an effective (often distance dependent) dielectric constant into the Coulomb potential,[17-19] but solvent fluctuations have sometimes been incorporated as well.[16,20]

More to the point of this review, there has recently been progress in using Eq. (2.7) with microscopic theories. Hall and Wolynes[21] showed how time-dependent Hartree theory could be used to study how the electron would be affected by a polarizable solvent composed of discrete atoms. Simulation methods have also aided this endeavor. Sprik et al.[22] used path-integral Monte Carlo techniques to look at the ionization of an Li atom in $NH_{3(l)}$ and Dobrosavljevic et al.[23] were able to investigate the effects of liquid-state disorder on electronic structure with a diffusion Monte Carlo simulation of an atom in a hard-sphere liquid. In this last instance, the idea was to include the fact that the inner cores of the solvent atoms would repel a solute electron, both because of electron–electron repulsion and Pauli exclusion. Assuming the nucleus is at r_0, this model would make the coupling potentials in Eq. (2.4) equal to

$$v(\vec{r}_{0j}; \vec{q}) = \begin{cases} \infty & |\vec{r}_{0j} - \vec{q}| < d \\ 0 & |\vec{r}_{0j} - \vec{q}| > d \end{cases}$$

for some electron-solvent (pseudopotential) radius d.

Somewhat less literal, but even more rewarding, has been the concerted effort to model polarizability with intraatomic *Drude oscillators*.[24-37] If one takes the internal Hamiltonian for the jth atom to be of the Drude form,

$$\hat{H}_{0j} = \frac{1}{2}\alpha_0\omega^2\vec{p}_j^2 + \frac{1}{2}\alpha_0^{-1}\vec{q}_j^2, \tag{2.8}$$

where α_0 is the gas-phase polarizability of the atom and $\hbar\omega_0$ is a mean gas-phase excitation energy, there is a natural association of $\vec{q}_j$ with the instantaneous fluctuating dipole of the atomic charge cloud and of $\vec{p}_j$ with its conjugate momentum. These dipoles then interact by the standard point–dipole point–dipole interaction, Eq. (2.6), with the tensor

$$\mathbf{T}(\vec{r}) = (1 - 3\hat{r}\hat{r})/r^3. \tag{2.9}$$

The rewards from this model have been impressive. At the classical level,[24,25] the way in which the condensed-phase polarizability α is renormalized from its gas-phase value follows immediately from the identification with the ensemble average

$$\alpha = (3k_BT)^{-1}\langle\vec{q}^{\,2}\rangle, \tag{2.10}$$

When an extension to quantum mechanics is made via path integrals, the moment on the right-hand side of Eq. (2.10) is seen to be just a special case of the imaginary-time correlation function $\langle qq(\tau)\rangle$, a quantity which carries all the information about the electronic absorption spectrum of the liquid. As has been shown in the literature, this realization suffices to create nontrivial theories for solvent effects on UV-visible spectra[27,29,30,37] and for excitons in liquids.[31] Moreover, it is relatively easy to generalize the internal Hamiltonian and the interactions so as to include quadrupoles[32-34]—leading to a theory for collision-induced spectra in liquids.[35]

B. Discrete Internal Coordinates

It is sometimes advantageous to think of intramolecular states as discrete possibilities—perhaps a conformation can either be *gauche* or *anti*, or an atom can only be in a finite number of electronic states. In such cases the intramolecular coordinates are discrete variables. What makes these models so useful, aside from their conceptual simplicity, is that the algebra of discrete variables is enormously helpful in formulating the problem. In fact, it turns out that *any* discrete model can be formulated in much the same way.

Consider a collection of atoms with m classical internal states, labeled by $\alpha = (1,\dots,m)$. The instantaneous state of an atom is then specified by a set of

m occupation numbers, v_α, each defined to be 1 if the atom is in state α, and 0 otherwise. Thus the role of the internal coordinate q_j is now played by

$$\vec{v}_j \equiv \{v_{j\alpha}; \alpha = 1, \ldots, m\}, \qquad (2.11)$$

with each $v_{j\alpha}$ equal to either 0 or 1 and

$$\sum_{\alpha=1}^{m} v_{j\alpha} = 1. \qquad (2.12)$$

Because $v_{j\alpha}$ raised to any power is $v_{j\alpha}$, *any* function of the vectors $\vec{v}_j$ and $\vec{v}_k$ can be written as a linear function of the components, so the interaction potentials in Eq. (2.4) can always be written

$$v(r_{jk}; \vec{v}_j, \vec{v}_k) = \sum_{\alpha, \beta} v_{j\alpha} A_{\alpha\beta}(r_{jk}) v_{k\beta},$$

or, in matrix notation,

$$v(r_{jk}; \vec{v}_j, \vec{v}_k) = \vec{v}_j \cdot A(r_{jk}) \cdot \vec{v}_k, \qquad (2.13)$$

with A a symmetric $m \times m$ matrix. Similarly, the isolated atom Hamiltonian can always be written as

$$\hat{H}_{0j} = \sum_{\alpha} h_\alpha v_{j\alpha} = \vec{h} \cdot \vec{v}_j, \qquad (2.14)$$

with h_α the gas-phase internal energy associated with an atom in its αth state.

By way of example,[38] to describe a solution or a molten salt containing Cu ions which can exist in either a $+1$ or $+2$ oxidation state, one could take a two-state model with $\alpha, \beta = (1, 2)$. The four possible Cu–Cu Coulomb's law interactions would be summarized neatly by writing

$$A_{\alpha\beta}(r) = e^2(\alpha)(\beta)/r.$$

The internal Hamiltonian would then be prescribed by the requirement that h_α equal the ionization potential of $Cu^{+\alpha}$. A completely analogous formulation could be used to describe the situation in certain liquid–metal alloys (such as CsAu), where the oxidation states of the species are thought to vary strongly with thermodynamic conditions.[39]

Another kind of example is provided by a simple model for *electron correlation* in a liquid of hydrogenic atoms.[5] Suppose we take as a basis the four sp^3 hybrid orbitals made from the valence orbitals on each atom. If we were to make the (classical mechanical) assumption that the electronic states of the atoms were defined by which hybrid orbital the valence electron occupied ($\alpha = 1, \ldots, 4$), then the instantaneous electronic structure of the liquid would be specified by the set of occupation numbers $\vec{v}_j$. However, hybrid orbitals have dipole moments, so each atom j would then have an internal-state-dependent dipole moment

$$\vec{\mu}_j = \mu_0 \sum_\alpha v_{j\alpha} \hat{\alpha}, \qquad (2.15)$$

with μ_0 the magnitude of a hybrid-orbital dipole moment and $\hat{\alpha}$ a unit vector parallel to the αth hybrid orbital.

The physical idea is that at densities sufficiently low that electron exchange between atoms is negligible (as in expanded liquid metals) most of the electron–electron correlation is due to the interaction of these fluctuating dipoles.[5,40–43] That is, the pair interactions are given by Eqs. (2.6) and (2.9) with $\vec{q}_j = \vec{\mu}_j$. Hence all we would need to flesh out the model is an internal Hamiltonian.

Of course, a classical description of the electronic structure is hardly likely to be adequate, but as we have alluded to already, and will discuss in more detail later, what one has to make quantum mechanical in these kinds of situations is the purely intramolecular part—so the interaction we have been describing may be quite reasonable. The internal Hamiltonian, though, must reflect the fact that the sp^3 hybrids are not eigenstates of the isolated atoms; the s and p orbitals are. The H_0 must therefore be written as an off-diagonal matrix in the sp^3 basis

$$\hat{H}_{0j} = -\frac{1}{4} \Delta E_{sp} \mathbf{h},$$

$$(\mathbf{h})_{\alpha\beta} \equiv h_{\alpha\beta} = 1 - 2\delta_{\alpha\beta}, \qquad (2.16)$$

where ΔE_{sp} is the gas-phase s-p promotion energy.

It may be somewhat easier to understand this model if we translate it into a second-quantized language. If we let $a_{j\alpha}^+$ and $a_{j\alpha}$ be the operators that create and annihilate (respectively) an electron in the αth hybrid orbital on the jth atom, making $\hat{n}_{j\alpha} = a_{j\alpha}^+ a_{j\alpha}$ the number operator for that orbital, then the single-atom Hamiltonian is

$$\hat{H}_{0j} = \sum_{\alpha, \beta} a_{j\alpha}^+ h_{\alpha\beta} a_{j\beta}, \qquad (2.17)$$

and the interparticle interaction is precisely as we described it above, except that $\hat{n}_{j\alpha}$ replaces $v_{j\alpha}$ in Eq. (2.15). In other words, the electron correlation is due to the interaction

$$V_I = \mu_0^2 \sum_{j<k} \hat{\vec{n}}_j \cdot T(r_{jk}) \cdot \hat{\vec{n}}_k, \qquad (2.18)$$

provided we make the obvious definition of a number operator vector

$$\hat{\vec{n}}_j = \sum_\alpha \hat{n}_{j\alpha} \hat{\alpha}.$$

By comparison, the more standard Hubbard model for electron correlation includes electron–electron interaction only when the electrons are on the same atom.[44]

One of the virtues of the occupation number representation of intramolecular dynamics is that this same kind of translation into second-quantized form is almost always this easy. The occupation numbers v_j are in fact the eigenvalues of the number operators $\hat{n}_j$, so the classical and quantal interactions are basically indistinguishable. Yet another intriguing feature is that the resulting models are automatically equivalent to classical (or quantum) spin models—but models situated in a liquid rather than a lattice.[38,45−56] Equivalently, one could think of these as spin models with fluctuating coupling constants.

To be concrete, suppose that we had atoms with m possible internal states that had an interaction $v_0(r)$ if the atoms were in the same state and $v_1(r)$ otherwise. The pair potential, which can be expressed in terms of occupation numbers as

$$\begin{aligned} v(r_{jk}; \vec{v}_j, \vec{v}_k) &= [v_0(r_{jk}) - v_1(r_{jk})] \sum_\alpha v_{j\alpha} v_{k\alpha} + v_1(r_{jk}) \\ &= [v_0(r_{jk}) - v_1(r_{jk})]\vec{v}_j \cdot \vec{v}_k + v_1(r_{jk}), \qquad (2.19) \end{aligned}$$

is precisely the interaction of the type of lattice model called a *Potts* model.[57] In the special case that $m = 2$ we are describing an *Ising* model.

An obvious reason that these connections are interesting is because the underlying lattice models have well-studied phase transitions of particular orders and symmetries. Thus the liquid itself may display some sort of transition. On a somewhat deeper level, the analogies are revealing because liquids with internal spin states combine the features of both disordered lattice problems and simple liquids. One should in principle therefore be able to use these models to probe the relationship between excitations in liquids and amorphous solids.

We close this section with one more illustration of a finite-state quantum mechanical problem, this one concerning *tunneling in liquids*.[38,51−54] Given that NH_3 molecules can tunnel between two pyramidal configurations that differ mainly in the direction of thier dipole moments, an (exceedingly) simple picture of fluid NH_3 might imagine a two-state description of ammonia molecules in which like states attract and unlike states repel. Such an interaction, however, is nothing but the Ising case ($m = 2$) of Eq. (2.19), so our system is that of an Ising model in a liquid. (Dipoles, of course, only have this behavior for certain orientations with respect to the line of centers. As we have discussed, it would be easy to replace this potential with an honest tensor interaction if desired.)

Still, as with the electron-correlation case, the states that form our basis are not eigenstates of the single-atom Hamiltonian. In the traditional double-well picture, the true eigenstates are symmetric and antisymmetric linear combinations of the "left-well" and "right-well" states we have been using. So, if we call K the matrix element of the isolated-molecule Hamiltonian between the $\alpha = 1$ and $\alpha = 2$ (left and right) states, then we need to write

$$\hat{H}_{0j} = -K\,\mathbf{h}, \tag{2.20}$$

where, now

$$(\mathbf{h})_{\alpha\beta} = 1 - \delta_{\alpha\beta}, \tag{2.21}$$

is the Pauli σ_x matrix.

In physical terms, this K is proportional to the tunneling rate between the two states, but it has yet another interpretation in the spin language. Our entire Hamiltonian can be expressed in terms of 2×2 Pauli matrices if spin states in each atom are written as the spinors

$$\begin{pmatrix} 1 \\ 0 \end{pmatrix}, \alpha = 1 \quad \text{or} \quad \begin{pmatrix} 0 \\ 1 \end{pmatrix}, \alpha = 2,$$

simply by saying

$$\hat{H} = \sum_{j<k} v(r_{jk})\sigma_{jz}\sigma_{kz} - \sum_j K\sigma_{jx}, \tag{2.22}$$

with $v = v_0 = -v_1$ and

$$\sigma_{jz} = \begin{pmatrix} 1 & 0 \\ 0 & -1 \end{pmatrix} \quad \sigma_{jx} = \begin{pmatrix} 0 & 1 \\ 1 & 0 \end{pmatrix}$$

defined to operate only on the spin in the jth atom. Since the average of σ_z defines magnetization, the K acts as a magnetic field perpendicular to direction of magnetization. Equation (2.22) is therefore said to be an *Ising model in a transverse field,*[58] a model that has long been used to describe tunneling-induced phase transitions in solids.[59]

C. Artificial Degrees of Freedom

Among the more intriguing developments of recent years have been liquid models in which one introduces completely artificial internal degrees of freedom for the purpose of making the resulting liquid problem mathematically equivalent to some other problem. The internal coordinate that each molecule carries is not meant to have any direct physical significance, but the partition function of a liquid augmented in this way can act as a generating function for a variety of physical phenomena.

It is, for example, possible to turn the quantum mechanical problem of calculating the *band structure of a liquid* into a purely classical calculation for a liquid with vector internal coordinates.[60,61] Let us assume that the band structure is determined by a tight-binding model which places a single s orbital on every atom. That is, given that $t(r_{jk})$ is the Hamiltonian matrix element governing electron hopping from an atom j to an atom k, we assume that the full Hamiltonian can be written in second quantized form as

$$\hat{H} = \sum_{j \neq k} t(r_{jk}) a_j^+ a_k. \tag{2.23}$$

For any one liquid configuration $R = \{\vec{r}, \ldots, \vec{r}_N\}$, the set of N eigenvalues of this Hamiltonian are the ingredients in the instantaneous density of states

$$D_R(E) = N^{-1} \sum_{m=1}^{N} \delta[E - E_m(R)],$$

and the average of this quantity over all the liquid configurations is what yields the experimental density of states at an energy E

$$D(E) = \langle D_R(E) \rangle.$$

So, the problem of calculating the density of states inside the band is that of diagonalizing the $N \times N$ matrix $\mathbf{H}(R)$ prescribed by Eq. (2.23) and then averaging the results over the liquid structure.

The first insight that makes this process possible is that the density of states can be determined directly from the Green's function[62]

$$G_R(E) = Tr[E\mathbf{1} - \mathbf{H}(R)]^{-1},$$

which, in turn, can be expressed as an integral using the matrix identity

$$Tr\,\mathbf{M}^{-1} = \frac{\int_{-\infty}^{\infty} dx_1 \ldots \int_{-\infty}^{\infty} dx_N 2i \sum_{j=1}^{N} x_j^2 \exp\{-i\sum_{j,k} x_j x_k \mathbf{M}_{jk}\}}{\int_{-\infty}^{\infty} dx_1 \ldots \int_{-\infty}^{\infty} dx_N \exp\{-i\sum_{j,k} x_j x_k \mathbf{M}_{jk}\}}$$

which is valid for any invertible matrix $\mathbf{M}$.

In our case, the matrix elements depend on the liquid configuration, so this ratio has to be averaged. We can do so by noting that the so-called replica identity, $A/B = \lim_{n\to 0} AB^{n-1}$, implies that the ratio of any two integrals can be written as a multiple integral over a vector coordinate:

$$\left[\int dx\, f(x)g(x)\right] \Big/ \left[\int dx\, g(x)\right]$$

$$= \lim_{n\to 0} \left[\int dx\, f(x)g(x)\right]\left[\int dx\, g(x)\right]^{n-1}$$

$$= \lim_{n\to 0} \int dx^{(n)} f(x^{(n)})g(x^{(n)}) \int dx^{(1)} \ldots \int dx^{(n-1)} g(x^{(1)}) \ldots g(x^{(n-1)})$$

$$= \lim_{n\to 0} \int dx\, f(x^{(a)}) \prod_{b=1}^{n} g(x^{(b)}).$$

If we associate one such vector coordinate with every atom, the liquid average over the Boltzmann factor $\exp\{-\beta V_B(R)\}$ (with V_B taken from Eq. 2.3) can then be performed directly to yield the averaged Green's function

$$\langle G_R(E)\rangle = \lim_{n\to 0} Q_B^{-1} \int d\vec{r}_1 \int dx_1 \ldots \int d\vec{r}_N \int dx_N 2iN(x_j^{(a)})^2$$

$$\prod_{j=1}^{N} \rho_0(x_j) \exp\left\{-\beta \sum_{j<k} u(r_{jk}; x_j, x_k)\right\},$$

where $u(r_{jk}; x_j, x_k)$ is an effective pair potential

$$-\beta u(r_{jk}; x_j, x_k) = -\beta u_0(r_{jk}) + 2ix_j \cdot x_k t(r_{jk}),$$

the $\rho_0(x_j)$ are unnormalized gas-phase distribution functions

$$\rho_0(x) = \exp\{-iEx \cdot x\},$$

and Q_B is the partition function for the liquid.

What this rather roundabout sequence of mathematical manipulations demonstrates is that we can indeed find the desired density of states in our liquid from a classical treatment. If we introduce an artificial, fluctuating, internal coordinate x on every atom, and then let the atom interact via effective pair potentials, the Green's function comes right out of the average of any one component of x

$$\langle G_R(E) \rangle = \lim_{n \to 0} 2iN \langle (x_j^{(a)})^2 \rangle.$$

Thus the statistical mechanics of our effective liquid completely determines the density of states.

Artificial internal states can also be used to study *classical percolation* in liquids.[63-66] Again, suppose we consider a simple illustration. In continuum percolation problems[67] one is interested in the connectedness function $g^\ddagger(r)$, the probability density for two particles to be a distance r away from each other but, nonetheless, at least indirectly *connected* to one another.[68] To be specific, define two particles to be directly connected to each other if they are within a distance d, and indirectly connected if there is a chain of connected particles stretching between them. Percolation would occur for such a system when the average number of atoms connected to a given atom became macroscopic, that is, when

$$\rho \int d\vec{r}\, g^\ddagger(r)$$

diverged.

This connectedness function can be found by looking at none other than an m-state Potts model in a liquid, Eq. (2.19). To implement our definition of connectivity in Eq. (2.19), take $v_0(r)$ to be zero and $v_1(r)$ to be given by

$$v_1(r) = \begin{cases} \infty & r < d \\ 0 & r > d. \end{cases}$$

Now construct an effective liquid with our standard Hamiltonian, Eqs. (2.1)–(2.4). The claim is that percolation in a liquid with potential V_B can be studied by examining the radial distribution functions of the effective liquid in the limit that the number of internal states goes to 1

$$g^\ddagger(r) = \lim_{m \to 1} [g(r; \alpha, \alpha) - g(r; \alpha, \beta)].$$

Here $g(r; \alpha, \alpha)$ and $g(r; \alpha, \beta)$ are the radial distribution functions for atoms that are in the same and in different internal states (respectively). (Distribution functions for internal degree of freedom problems are discussed in Section IV.A.)

We shall not attempt to prove this statement, but it is easy enough to see why it holds. When two particles are connected, the pair potential $v(r_{jk}; \vec{v}_j, \vec{v}_k)$ is infinite unless the atoms are in the same internal state. Hence, at finite temperatures, all the particles that are connected in any one cluster in the effective liquid must be in the same internal state. But given this fact, $\lim_{m \to 1} g(r; \alpha, \beta)$ can be shown to be the disconnectedness (or blocking) function—the radial distribution function for explicitly *unconnected* atoms. Since $\lim_{m \to 1} g(r; \alpha, \alpha)$ is just the (ordinary) radial distribution function for the original liquid, the difference of the two functions is the desired connectedness function.

III. THE ROLE OF QUANTUM MECHANICS

A. Qualitative Considerations

It goes without saying that an exact quantum mechanical treatment of statistical mechanical problems is a desirable goal. Whenever we can solve the Schrödinger equation and do the appropriate averaging for every electron and nucleus in our system we have every right to feel proud of ourselves. In liquid problems, however, this goal is usually worse than unacheivable—it actually disguises the reality of liquids. With the notable exceptions of superfluids and Fermi liquids, center-of-mass translation is fundamentally classical.

One can see this fact in a number of ways. The simplest is note that the deBroglie wavelength of a molecule with mass M

$$\Lambda = (2\pi\hbar^2\beta/M)^{1/2},$$

and $\beta = (k_B T)^{-1}$, is only of the order of 0.15 Å for a molecule at room temperature with a molecular weight of 50 amu, a figure 20–50 times smaller than typical molecular sizes. Besides, even when such translational quantum effects become noticeable—such as in He and Ne below 20 K—the main consequence is that the particles act as if they were interacting under a slightly softened classical pair potential[69, 70] (an effect which, admittedly, seems to have interesting ramifications for freezing of quantum solids).[71]

If we take the classical character of simple liquids for granted, the question for us is what the implications are for intramolecular behavior—which is certainly *not* going to be especially classical. The issue is made more interesting by the fact that the quantal dynamics inside any one molecule is likely to be affected by the equally quantal dynamics inside the surrounding moleucles. Thus, the environment a molecule sees in a neat liquid will have both classical and quantum features. To start our analysis though, let us consider the simpler case relevant to solution spectroscopy: a single solute with quantum mechanical internal dynamics dissolved in a liquid of featureless solvent molecules.

The Hamiltonian for this kind of system can be written as

$$\hat{H}(q, R) = \hat{H}_0(q) + V_I(q, R) + V_B(R), \tag{3.1}$$

where R is the set of solvent position vectors, as in the previous section, but now q is a single, at most few-dimensional, coordinate affiliated with one atom. In the *classical* limit, the probability density governing the distribution of q in the liquid is

$$s(q) = Q^{-1} \int dR \exp\{-\beta H(q, R)\} \tag{3.2}$$

$$s(q) = Q^{-1} \exp\{-\beta H_0(q)\} \int dR \exp\{-\beta[V_I(q, R) + V_B(R)]\} \tag{3.3}$$

with Q the partition function for the system

$$Q = \int dR \int dq \exp\{-\beta H(q, R)\}. \tag{3.4}$$

(Technically, Q is the configurational integral since the factor due to the momentum integration is being ignored.) One can always imagine formally integrating out the solvent coordinates in Eq. (3.3), leaving something of the form

$$s(q) = Q^{-1} \exp\{-\beta[H_0(q) + W(q)]\}. \tag{3.5}$$

The function $V_{MF}(q) = H_0(q) + W(q)$ can be interpreted as the average potential surface on which the coordinate q moves in the solution—which makes $W(q)$ the average correction due to the solvent effects. We therefore call V_{MF} the *potential of mean force* for the internal coordinate.

Now let us try to repeat this calculation, bringing in the fact that the internal dynamics is quantal but the bath behavior is not. Superficially all one has to do to modify Eq. (3.3) is to replace H_0 with $\hat{H}_0$ and the integral over q with a trace. The order of the operations now is crucial, however. If the integral over the bath is performed first, one is left with the quantum analogue of Eq. (3.5). That is, the probability density for q would come directly out of the wave functions and energy levels of the potential of mean force. On the other hand, if the quantum mechanics of the internal coordinate were determined by solving the Schrödinger equation for the q-dependent part

$$\hat{H}_{ad}(q; R) = \hat{H}_0(q) + V_I(q, R), \tag{3.6}$$

16 RICHARD M. STRATT

for each set of solvent positions R, and *then* we did the integral over the bath, there would be considerably more effort involved.

Unfortunately it is fairly easy to convince oneself that the second procedure, which we shall call the *adiabatic approximation*,[3,4] is usually going to be significantly more accurate than the first procedure, what we might call the *sudden approximation*. The bath coordinates are expected to be slower (have lower frequencies and higher masses) than the internal coordinate—indeed that is what one usually means by being more classical. So just as in the Born–Oppenheimer treatment of fast electrons in the presence of slow nuclei, the optimum approach is to start with the quantum mechanics for fixed values of the slow variable (start with the electronic structure for fixed molecular geometries). The energy levels parameterized by the slow coordinates then define potential surfaces on which the slow variable dynamics (such as vibration) can be studied.[72]

This adiabatic approach is formally equivalent to assuming that the bath kinetic energy commutes with the rest of the Hamiltonian—an exact statement in the limit that the mass of the bath particles is much larger than that of the internal coordinate. The approximation of quantizing on a potential of mean force, though, seems to be assuming that the bath is sufficiently *faster* than the internal coordinate that the motion of the bath atoms can be taken as instantaneous on the intramolecular time scale—hence our use of the term "sudden."

We can get some insight into the relative merits of these two schemes with the aid of a simple model. Consider two coupled harmonic oscillators, one of which we shall call the bath (subscript B) and the other the internal coordinate (subscript 0). If we let them be linearly coupled with a coupling strength α, the Hamiltonian becomes

$$\hat{H} = \frac{\hat{p}_0^2}{2m_0} + \frac{\hat{p}_B^2}{2m_B} + \frac{1}{2}m_0\omega_0^2 x_0^2 + \frac{1}{2}m_B\omega_B^2 x_B^2 + \alpha x_0 x_B.$$

The model is characterized by two dimensionless parameters,

$$r = (\omega_B/\omega_0)^2,$$

which compares the bath frequency to the intramolecular frequency, and

$$u = (\alpha^2/m_0 m_B \omega_0^2 \omega_B^2),$$

which is a measure of the coupling strength.

The exact quantum mechanical partition function for this model can be evaluated immediately by a normal mode transformation, leading to a product of two, quantal, simple harmonic oscillator partition functions:

$$Q = [4\sinh(\hbar\omega_+\beta/2)\sinh(\hbar\omega_-\beta/2)]^{-1},$$

$$\omega_\pm^2 \equiv \tfrac{1}{2}\omega_0^2\{1 + r \pm [1 + 2(2u - 1)r + r^2]^{1/2}\}.$$

However, we can also calculate the partition function in the adiabatic approximation Q_{ad} by doing the quantum mechanics for fixed x_B and then integrating over x_B. This sequence yields the product of the gas-phase quantal partition function for the oscillator and a classical partition function for the bath

$$Q_{ad} = [2\sinh(\hbar\omega_0\beta/2)(\hbar\omega_{eff}^{bath}\beta)]^{-1},$$

with an effective bath frequency

$$(\omega_{eff}^{bath})^2 = \omega_B^2(1 - u).$$

Similarly, we can also compute the sudden partition function Q_{sud} by integrating over x_B first and then doing quantum mechanics on what remains. The end result then is the product of the classical partition function for the uncoupled bath with a quantal partition function for the oscillator

$$Q_{sud} = [2\sinh(\hbar\omega_{eff}^{solute}\beta/2)(\hbar\omega_B\beta)]^{-1},$$

at an effective frequency

$$(\omega_{eff}^{solute})^2 = \omega_0^2(1 - u).$$

One way to compare these formulas is to try to develop expansions of the exact result that give, to leading order, the two different approximate results. After doing so it is straightforward to verify that

$$Q = Q_{ad}[1 + \text{terms of order } (r)(u)].$$

In other words, the adiabatic formula becomes correct in *either* the limit that the bath is much slower than the solute ($r \ll 1$) *or* the limit that the coupling is weak ($u \ll 1$). The sudden result, however, only reduces to the exact formula if the coupling is weak—and there is no reason to expact to be in the weak coupling regime just because the internal and bath time scales differ.

These differences in behavior are illustrated in Fig. 1, where we have plotted the solvation free energy for the model (the excess free energy of the coupled

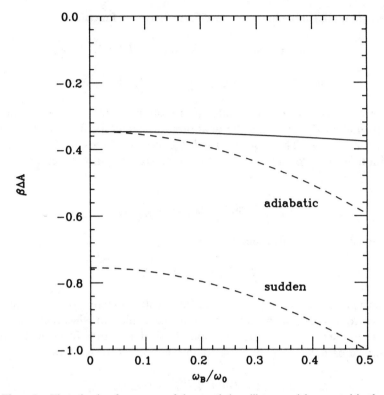

Figure 1. The solvation free energy of the coupled-oscillator model presented in the text (in units of $k_B T$) as a function of the ratio of the bath frequency ω_B to the intramolecular frequency ω_0. The solid line is the exact result and the two dashed lines are approximations. We have taken $\hbar\omega_0 = 5k_B T$ and the coupling strength $u = 0.5$.

system over that of the separated oscillators)

$$\beta\Delta A = -\ln(Q/Q_0 Q_B).$$

Here, Q_0 and Q_B are the exact quantum mechanical partition functions for the internal coordinate and the bath (respectively). Of course, if both of the oscillators were taken to be classical, all three forms of the partition function would be identical. The order in which the integrals are done in classical statistical mechanics is irrelevant.

Much the same set of conclusions would be reached on analyzing the case of many interacting molecules with quantal internal degrees of freedom. If the molecular translation is to be regarded as being classical, it is important to make an adiabatic, and not a sudden, approximation. From here on out we assume that we are doing so, without further comment.

B. Discretized Path Integral Formulation

Even with an adiabatic approximation, the statistical mechanics of quantal internal degrees of freedom in a liquid is still a quantum mechanical problem. It might therefore seem that one would have to set up a fully quantum mechanical many-body formalism to understand these systems. The fact that the hardest part of the problem—the averaging over the complicated many-particle position correlations—is fundamentally a classical exercise (and a *solved* exercise at that) would seem to be of no particular help.

This expectation is, not surprisingly, not correct. One should always be able to embody the physics of the situation in the mathematics; it is just a matter of finding the appropriate methods. As we shall show in this section, representing the quantal degrees of freedom with the aid of *discretized Feynman path integrals* will turn any problem of the form we have been considering into a purely classical statistical mechanical calculation.[73,74] We can thus make use of all of the classical techniques and physical insights into simple liquids that have been gathered over the years.[3,75,76]

Let us start with the partition function for a neat liquid of atoms with quantal internal degrees of freedom.[52] In general the partition function can be written as the trace of the Boltzmann density matrix[74] for our Hamiltonian $\hat{H}$

$$Q = \int dR \int dq \, \langle R, q | e^{-\beta \hat{H}} | R, q \rangle, \tag{3.7}$$

where by the matrix element we mean the sum over all the exact eigenstates n of the whole system

$$\langle R, q | e^{-\beta \hat{H}} | R, q \rangle = \sum_n \psi_n^*(R, q) \psi_n(R, q) \exp\{-\beta E_n\}, \tag{3.8}$$

and ψ_n and E_n are the wave functions and energy levels, respectively. We now make the adiabatic approximation by replacing the wave functions and energies in this matrix element by the eigenfunctions and eigenvalues of $\hat{H}_{ad}$, Eq. (3.6):

$$\langle R, q | e^{-\beta \hat{H}} | R, q \rangle \approx e^{-\beta V_B(R)} \sum_n \psi_n^*(q; R) \psi_n(q; R) \exp\{-\beta E_n(R)\}, \tag{3.9}$$

This discretized path integral is generated by inserting P complete sets of eigenstates of $\hat{H}_{ad}$ into Eq. (3.7). Since these states satisfy the relation

$$\int dq \, |q; R\rangle \langle q; R| = 1,$$

we can introduce P copies of the internal coordinates

$$Q = \int dR \int dq^{(1)} \cdots \int dq^{(P)} e^{-\beta V_B(R)} \prod_{t=1}^{P} \langle q^{(t)}; R | e^{-\beta \hat{H}_{ad}/P} | q^{(t+1)}; R \rangle,$$

$$q^{(P+1)} \equiv q^{(1)}, \tag{3.10}$$

without making any further approximations. However, in the limit of large P, the exponential of the sum of operators that comprises the Boltzmann factor for $\hat{H}_{ad}$ becomes equal to the product of the exponentials of the operators (the Trotter product formula).[77] That is,

$$\lim_{P \to \infty} \langle q^{(t)}; R | e^{-\beta \hat{H}_{ad}/P} | q^{(t+1)}; R \rangle = \langle q^{(t)} | e^{-\beta \hat{H}_0/P} | q^{(t+1)} \rangle e^{-(\beta/P)V_I(q(t), R)}. \tag{3.11}$$

Notice how the assumption that V_I is *not* an operator allows us to bring a Boltzmann factor out from inside the q matrix element.

The symbol q refers to all N of the internal coordinates in the liquid, so any one wave function $\langle q^{(t)} |$ is actually a product of N independent wave functions, one per atom

$$\langle q^{(t)} | \equiv \prod_{j=1}^{N} \langle q_j^{(t)} |.$$

Thus, from Eq. (2.2),

$$\langle q^{(t)} | e^{-\beta \hat{H}_0/P} | q^{(t+1)} \rangle = \prod_{j=1}^{N} \langle q_j^{(t)} | e^{-\beta \hat{H}_{0j}/P} | q_j^{(t+1)} \rangle$$

$$= \prod_{j=1}^{N} \rho_0(q_j^{(t)}, q_j^{(t+1)}; \beta/P), \tag{3.12}$$

where ρ_0 is the off-diagonal Boltzmann density matrix for the internal coordinate of any one isolated atom:

$$\rho_0(q_j^{(t)}, q_j^{(t+1)}; \beta) = \sum_n \psi_n^{0*}(q_j^{(t)}) \psi_n^0(q_j^{(t+1)}) \exp\{-\beta E_n^0\}, \tag{3.13}$$

if we take ψ_n^0 and E_n^0 to be eigenfunctions and eigenvalues of $\hat{H}_{0j}$.

We are now in a position to assemble the full partition function. Substituting Eqs. (2.3), (2.4), (3.11), and (3.12) into Eq. (3.10) yields

$$Q = \int dR \int dq^{(1)} \dots \int dq^{(P)} \prod_{t=1}^{P} \prod_{j=1}^{N} \rho_0(q_j^{(t)}, q_j^{(t+1)}; \beta/P)$$

$$\exp\left\{-\beta \sum_{j<k} u_0(r_{jk})\right\} \exp\left\{-(\beta/P) \sum_{t=1}^{P} \sum_{j<k} v(r_{jk}; q_j^{(t)}, q_k^{(t)})\right\},$$

a rather unwieldy expression. Nonetheless, a considerable simplification ensues if we group together the coordinates associated with each particle j instead of with each "time" t. Call the set of P internal coordinates $q_j^{(t)}$ affiliated with j the vector q_j. The gas-phase distribution function governing the probability of having a given q_j is just the density-matrix product

$$\rho_0(q_j) \equiv \prod_{t=1}^{P} \rho_0(q_j^{(t)}, q_j^{(t+1)}; \beta/P),$$

$$q^{(P+1)} \equiv q^{(1)}. \tag{3.14}$$

Since

$$\int dq^{(1)} \dots \int dq^{(P)} = \int dq_1 \dots \int dq_N,$$

the partition function can be written

$$Q = \int dR \int dq_1 \dots \int dq_N \prod_{j=1}^{N} \rho_0(q_j)$$

$$\times \exp\left\{-\beta \sum_{j<k} u_0(r_{jk})\right\} \exp\left\{-\beta \sum_{j<k} v(r_{jk}; q_j, q_k)\right\}, \tag{3.15}$$

with the effective potential

$$v(r_{jk}; q_j, q_k) \equiv (P)^{-1} \sum_{t=1}^{P} v(r_{jk}; q_j^{(t)}, q_k^{(t)}). \tag{3.16}$$

In other words, *the partition function of our system is precisely that of a classical liquid of atoms with (classical) vector internal degrees of freedom q. The internal coordinates are subject to a gas-phase distribution $\rho_0(q)$ and the* *atoms interact via effective pair potentials*

$$u(r_{jk}; q_j, q_k) \equiv u_0(r_{jk}) + v(r_{jk}; q_j, q_k). \qquad (3.17)$$

Equation (3.15), along with Eqs. (3.14) and (3.16), are really all that we need to study the problems posed in Section II, but since we did introduce P copies of our internal variables seemingly out of nowhere, it might be worthwhile to make some remarks about their significance. To begin with, we can remind ourselves of the connection with ordinary (nondiscretized) Feynman path integrals.[73,74] The set of variables $q^{(1)}$, $q^{(2)}$, ..., $q^{(P)}$, can be thought of as a sequence of values taken on by q as a time t evolves; in the limit that $P \to \infty$, we can actually think of a function $q(t)$. In that sense, q defines a *path*, and the integral over all possible intermediate stops on that path, dq, is an integral over all possible paths $Dq(t)$.

Most of the applications of Feynman path integrals have used this kind of continuum notation. Our effective potential, Eq. (3.16), for example, could be expressed as

$$\int d\tau \, v(r_{jk}; q_j(\tau), q_k(\tau)),$$

if we defined $\tau \equiv t/P$. This notation is particularly helpful in semiclassical applications. There, one picks out the most important path(s) by looking for stationary values of the effective Hamiltonian (the action). If the action is written as an integral, finding the optimimum paths is a textbook example of a calculus-of-variations problem, typically leading to a $q(t)$ satisfying some sort of classical equation of motion.[78] So far, though, this approach has not proven to be as convenient as the discretized formulation in intramolecular problems.

The discretized formulation, for one thing, lends itself to an interesting physical picture of the role of quantum mechanics. What one normally tries to express with a wave function $\psi_n(q)$ is the idea of Heisenberg uncertainty— that there is no definite value for a coordinate q, only a probability distribution. The discretized path integral scheme does not use wave functions; instead it introduces an infinite number of copies of q, each one of which can have a definite value. In a well-defined mathematical sense these approaches are identical, but the latter fits more naturally into a classical calculation. We shall find this analogy a helpful one in our subsequent applications.

Even more germane is the special form that the action takes on when discretization is used. Consider first the effective coupling potential, Eq. (3.16). In examples in which the coupling is bilinear, Eqs. (2.5) and (2.6), the fact that internal coordinates are only coupled at the same time t means that the effective interaction always involves a dot product of the internal-degree-of-freedom vectors q_j and q_k

$$v(r_{jk}; q_j, q_k) = q_j \cdot q_k v(r_{jk})/P,$$
$$v(r_{jk}; \vec{q}_j, \vec{q}_k) = P^{-1} \vec{q}_j \cdot T(r_{jk}) \cdot \vec{q}_k,$$

(3.18)

(where the dot products in the tensor case refers to a simultaneous sum over both Cartesian indices and times t). Most of the models in Section II—and *all* models with discrete internal coordinates—will therefore have coupling potentials of this form.

The gas-phase internal distribution, Eq. (3.14), need not have a simple appearance in vector notation, at least for continuous variables, though it often does. But discrete variables, here again, will always lend themselves to compact expressions in our vector language. The Boltzmann density matrix in the discrete variable case becomes

$$\rho_0(\vec{v}^{(t)}, \vec{v}^{(t+1)}; \beta/P) = \vec{v}^{(t)} \cdot (e^{-(\beta/P)H_0}) \cdot \vec{v}^{(t+1)}$$

(3.19)

if H_0 is the isolated-atom Hamiltonian matrix in the finite-state basis (e.g., Eqs. 2.16 and 2.20). This structure then has immediate ramifications for the internal distribution. Supposing there are m internal states, let the $m \times m$ matrix $\mathbf{B}$ be defined by

$$B(\alpha, \beta) \equiv (\mathbf{B})_{\alpha, \beta} \equiv (e^{-(\beta/P)H_0})_{\alpha, \beta}.$$

(3.20)

Then

$$\rho_0(\vec{v}^{(t)}, \vec{v}^{(t+1)}; \beta/P) = \sum_{\alpha, \beta} v_\alpha^{(t)} B(\alpha, \beta) v_\beta^{(t+1)}$$
$$= \exp\left\{ \sum_{\alpha, \beta} v_\alpha^{(t)} \ln B(\alpha, \beta) v_\beta^{(t+1)} \right\},$$

so that the isolated-atom internal distribution can always be written

$$\rho_0(\vec{v}) = \exp\{\vec{v} \cdot H \cdot \vec{v}\},$$

(3.21)

with the dot products referring to both a sum over internal states α and time t, and the $m \times m \times P \times P$ "matrix" H equal to

$$H_{\alpha, \beta}(t, t') = \tfrac{1}{2} \ln B(\alpha, \beta)(\delta_{t, t'+1} + \delta_{t+1, t'}),$$

(3.22)

(if, as usual, an index of $P + 1$ is interpreted as an index of 1).

These findings apply to any finite-state problems. However, under the not uncommon circumstances that

$$\ln B(\alpha, \beta) = c + \delta_{\alpha, \beta} a, \tag{3.23}$$

for some constants a and c, an even further simplification takes place. Using the conservation-of-probability relation Eq. (2.12) enables us to rewrite Eqs. (3.21) and (3.22) as

$$\rho_0(\vec{v}) = \exp\{\vec{v} \cdot \mathbf{H} \cdot \vec{v} + Pc\}, \tag{3.24}$$

where now $\mathbf{H}$ is only a $P \times P$ matrix (with no internal state indices)

$$\mathbf{H}(t, t') = \tfrac{1}{2} a(\delta_{t, t'+1} + \delta_{t+1, t'}). \tag{3.25}$$

Perhaps a few explicit examples for continuous and discrete variables are in order:

1. *Harmonic oscillator.* For a one-dimensional simple harmonic oscillator, with a Hamiltonian of the form

$$\hat{H} = \frac{\hat{p}^2}{2m} + \frac{1}{2} m\omega^2 q^2,$$

the off-diagonal Boltzmann density matrix is given by

$$\rho_0(q^{(t)}, q^{(t+1)}; \beta/P) = [(m\omega/h) \operatorname{csch} x]^{1/2}$$

$$\exp\{-(m\omega/2\hbar)\{[(q^{(t)})^2 + (q^{(t+1)})^2] \coth x - 2q^{(t)}q^{(t+1)} \operatorname{csch} x\}\}$$

with $x = \hbar\omega\beta/P$. The internal distribution can thus be written in terms of the $P \times P$ symmetric matrix $\mathbf{H}$

$$\rho_0(q) = \exp\{-q \cdot \mathbf{H} \cdot q + C\},$$
$$(\mathbf{H})_{t, t'} = (m\omega/2\hbar)[2 \coth x \delta_{t, t'} - \operatorname{csch} x(\delta_{t, t'+1} + \delta_{t+1, t'})],$$
$$C = \frac{P}{2} \ln[(m\omega/h) \operatorname{csch} x], \tag{3.26}$$
$$(t = P + 1) \equiv (t = 1).$$

Three-dimensional harmonic oscillators (such as Eq. 2.8) are separable into independent one-dimensional oscillators, so the three-dimensional equivalent is just a threefold product of Eq. (3.26)

$$\rho_0(\vec{q}) = \exp\{-\vec{q}\cdot\mathbf{H}\cdot\vec{q} + 3C\}. \tag{3.27}$$

2. *Two-state tunneling.*[52] Taking the Hamiltonian to be that of Eqs. (2.20) and (2.21) leads to a **B** matrix satisfying Eq. (3.23) with

$$\begin{aligned} a &= \ln\coth(\beta K/P), \\ c &= \ln\sinh(\beta K/P). \end{aligned} \tag{3.28}$$

These constants can be substituted into Eqs. (3.24) and (3.25) to get the desired gas-phase distribution.

3. *Electron correlation.*[5] Similarly, the Hamiltonian of Eq. (2.16) implies Eq. (3.23) with

$$\begin{aligned} a &= \ln[(e^x + 3e^{-x})/(e^x - e^{-x})], \\ c &= \ln\tfrac{1}{2}\sinh x \end{aligned} \tag{3.29}$$

and $x = (\beta\Delta E_{sp}/2P)$. Once again, these two numbers suffice to determine the gas-phase distribution.

IV. CORRELATION FUNCTIONS AND MOMENTS

To this point we have done little more than set up a number of intramolecular statistical mechanics problems. Section II defined the problems and Section III showed how all of these problems can be rewritten as though they were classical, despite the very important role that quantum mechanics plays. Presumably, we ought to be ready to arrive at some of the answers. Before we do so, however, we shall spend some time figuring out just what answers we want.

A. Correlation Functions Relevant to Internal Degrees of Freedom

The central issue in the behavior of internal degrees of freedom is to understand how the probability distribution function of an internal coordinate on a given atom in the liquid $s(q)$,

$$\begin{aligned} s(q_i) &= Q^{-1}\int dq_{\neq i}\int dR\exp\{-\beta H(q, R)\}, \\ Q &= \int dq\int dR\exp\{-\beta H(q, R)\}, \end{aligned} \tag{4.1}$$

differs from that of an isolated atom $s_0(q)$,

$$s_0(q_i) = Q_0^{-1} \exp\{-\beta H_{0i}(q_i)\},$$

$$Q_0 = \int dq_i \exp\{-\beta H_{0i}(q_i)\}. \tag{4.2}$$

[The notation $dq_{\neq i}$ here means to integrate over the internal coordinates q_j of all the molecules ($j = 1, \ldots, N$), except the ith.] Notice that these purely classical mechanical expressions are also completely correct for quantum mechanical cases if the problem is formulated in discretized path integral language—as we shall assume from here on out.

From a formal perspective we can always define a function $y(q)$, the *cavity distribution function*, that embodies whatever the modification is that the liquid makes,[79]

$$s(q) = s_0(q)y(q) \Big/ \int dq\, s_0(q)y(q). \tag{4.3}$$

This function takes its name from the fact that it represents what the distribution for q_i would be if there were no internal Hamiltonian H_{0i} in the ith atom, just a cavity. That is, to within an arbitrary normalization

$$y(q_i) = Q_{(-i)}^{-1} \int dq_{\neq i} \int dR$$

$$\times \exp\left\{-\beta \sum_{j\neq i} H_{0j}(q_j) - \beta \sum_{j<k} u_0(r_{jk}) - \beta \sum_{j<k} v(r_{jk}; q_j, q_k)\right\}, \tag{4.4}$$

if one assumes the standard Hamiltonian, Eqs. (2.1)–(2.4), and $Q_{(-i)}$ is defined to be what the partition function would have been had the ith atom had no internal coordinate.

Within this particular normalization, though, we see that $y(q)$ is the ratio of the numerator of Eq. (4.1) to $Q_{(-i)}$ times the numerator of Eq. (4.2). This ratio has an interesting physical interpretation: If we consider the hypothetical liquid in which the internal coordinate of the ith atom is held fixed at the value q_i, then the numerator of Eq. (4.1) is the partition function for this liquid, $Q(q_i)$, which means that it is also the Boltzmann factor for the free energy of that liquid, $\exp\{-\beta A(q_i)\}$. Similarly, the numerator of Eq. (4.2) times $Q_{(-i)}$ is the partition function $Q_0(q_i) = \exp\{-\beta A_0(q_i)\}$ for the hypothetical situation in which the ith atom has $q = q_i$, but this q_i does not interact with the remainder of the liquid. Hence, the overall ratio, the cavity distribution function, is

$$y(q_i) = \exp\{-\beta[A(q_i) - A_0(q_i)]\}$$
$$= \exp\{-\beta\Delta\mu(q_i)\}, \tag{4.5}$$

the Boltzmann factor for the *excess chemical potential* (*over and above the gas-phase value*) *required for an internal coordinate to have a value* q_i *in the liquid*. Ordinary thermodynamic arguments tell us that Helmholtz free energy differences are actually the reversible work involved in a given process. So, we will find ourselves calculating the reversible work necessary to achieve a given internal coordinate value when we finally get around to computing cavity distribution functions.[80]

Besides such *intra*molecular correlation functions, these problems suggest defining new *inter*molecular correlation functions. The analogue of the radial distribution function for featureless particles is the radial distribution function $g(r_{jk}; q_j, q_k)$, defined so that

$$s(q_j)g(r_{jk}; q_j, q_k)s(q_k) = Q^{-1}V^2 \int dq_{\neq j,k} \int dR_{\neq j,k} \exp\{-\beta H(q, R)\}, \tag{4.6}$$

where V is the volume and, as above, the $\neq j, k$ subscript means to integrate over the coordinates of all the atoms except j and k. In more physical terms, we could also say that

$$g(r_{jk}; q_j, q_k)s(q_k) = \frac{V^2 \int dq_{\neq j,k} \int dR_{\neq j,k} \exp\{-\beta H(q, R)\}}{\int dq_{\neq j} \int dR \exp\{-\beta H(q, R)\}},$$

is the probability density for finding an atom with internal coordinate value q_k a distance r_{jk} away from a given, tagged, atom with internal coordinate value q_j.

Thermodynamic quantities can be expressed in terms of this new function in much the same way that they were with $g(r)$. For example, the excess internal energy—the average (potential) energy of the liquid solely due to the interatomic interactions—is

$$\langle U/N \rangle = \frac{1}{2}\rho \int dq_1 s(q_1) \int dq_2 s(q_2) \int d\vec{r} u(r; q_1, q_2)g(r; q_1, q_2), \tag{4.7}$$

with ρ the number density and the total pair potential defined to be

$$u(r; q_1, q_2) \equiv u_0(r) + v(r; q_1, q_2). \tag{4.8}$$

It is hard to miss the similarity of formulas of this sort with those for mixtures. If we label the components in a mixture by Greek indices, the excess internal energy is

$$\langle U/N \rangle = \frac{1}{2} \rho \sum_{\alpha, \beta} x_\alpha x_\beta \int dr\, u_{\alpha\beta}(r) g_{\alpha\beta}(r).$$

Here the role of $g(r; q_1, q_2)$ is played by $g_{\alpha\beta}(r)$, the radial distribution function between species α and β (and analogously for the pair potential u), and the intramolecular distribution functions $s(q)$ are replaced by mole fractions x_α. We shall find ourselves relying quite heavily on this interchangeability between results for mixtures and for internal degrees of freedom. It should be pointed out, though, that this relationship is not entirely an obvious one. In the mixture case, the relative proportions of the various components is fixed by some external constraint, whereas in our problems the internal state fluctuates according to the distribution $s(q)$. Indeed, in practice, one will not even know the distribution. It will have to be calculated from the properties of the liquid—which, in turn, will depend on the distribution itself. Thus there is always an extra *self-consistent* calculation required in internal-degree-of-freedom problems.[79]

The mixture-internal coordinate analogy, of course, has its fundamental origins in the relationship between the canonical and grand canonical ensembles.[81] The ensemble in which the mole fractions x_α are fixed and the one in which they fluctuate about some average differ only by a Legendre transform. Still, it was not clear just how simple this transform was to perform until a diagrammatic tour de force by Chandler and Pratt.[79,82] What they showed was that the very same Mayer cluster diagrams normally used to represent correlation functions for featureless single-component liquids represented liquids with internal coordinates equally well, provided one reinterpreted the vertices in the diagrams.

Let us be a little more specific using radial distribution functions as an illustration of this result. A useful diagrammatic series for the $g(r_{12})$ of a simple pure liquid is shown in Fig. 2. Here a line (bond) between any two vertices j and k symbolizes a Mayer f-bond between atoms j and k

$$f(r_{jk}) = \exp\{-\beta u(r_{jk})\} - 1,$$

with $u(r)$ the pair potential. The open (white) circles are assigned a value of unity and a white circle labeled 1 means that atom 1 is fixed at some position r_1 (likewise for one labeled 2). Each filled-in (black) circle (also called a vertex) is similar in that it means to imagine an arbitrary but different atom j at position r_j, but now one must integrate over all such positions, weighted by the probability density for finding an atom at that position. In a homogeneous

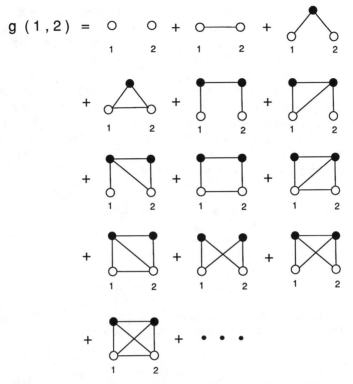

Figure 2.   Diagrammatic expansion of the radial distribution function. With appropriate definitions of the vertices and bonds, the same series applies to fluids with and without internal degrees of freedom. When internal degrees of freedom are present, the series applies to both classical and quantum mechanical cases.

fluid of number density ρ, this probability density is just ρ itself, so a black circle implies

$$\rho \int dr_j.$$

Thus, the value of the first diagram is just the number 1 and the sixth diagram is equal to

$$\rho \int d\vec{r}_3 \rho \int d\vec{r}_4 f(r_{13}) f(r_{34}) f(r_{24}) f(r_{14})$$

A more concise restatement of all of this information would be that these diagrams have white 1-circles, black ρ-circles, and f-bonds. (There is also a symmetry-number factor associated with every diagram that we shall not discuss here, other than to note that all the diagrams in Fig. 2 have this factor equal to 1, except for the last three. Symmetry numbers are discussed in the standard presentations of liquid theory.)[1,2]

The Chandler–Pratt finding, which we might call the fundamental theorem of intramolecular statistical mechanics, is that precisely the same series represents $g(r_{12}, q_1, q_2)$ with the appropriate identifications. A white circle labeled 1 now fixes both the position r_1 and the internal coordinate q_1 of particle 1. Similarly, a black circle is affiliated with both an r_j and a q_j, with the integration over them subject to their joint probability density

$$\int d(j) \equiv \rho \int d\vec{r}_j \int dq_j\, s(q_j). \tag{4.9}$$

The f-bond is still defined from the pair potential as usual:

$$f(j, k) \equiv f(r_{jk}; q_j, q_k) = \exp\{-\beta u(r_{jk}; q_j, q_k)\} - 1. \tag{4.10}$$

Hence, for an intramolecular problem, the first diagram of Fig. 2 would still be equal to 1 and the sixth diagram would be given by

$$\int d(3) \int d(4) f(1, 3) f(3, 4) f(2, 4) f(1, 4).$$

In our previous language we would say that the diagrams have white 1-circles, black ρs-circles, and f-bonds.

B. Moments of Internal Coordinates

For any scalar intramolecular coordinate q, our discussion in Section IV.A makes clear that the value of the nth moment in the liquid is

$$\langle q^n \rangle = \int dq\, s(q) q^n. \tag{4.11}$$

Moments such as these will actually be important to us in a number of ways. For one thing, the full intramolecular distribution function $s(q)$ is not always needed in order to understand a given experiment. Frequently, a few moments of the distribution suffice. It will also turn out, in certain classes of theories, that $s(q)$ is completely prescribed by a small number of moments. In view of the fact that we always have to solve for $s(q)$ self-consistently, it is to our advantage to have the self-consistency condition a small set of coupled algebraic equations for these moments, rather than being a functional equation for $s(q)$ in terms of itself.

An example of just such a simplification is provided by a *liquid of breathing hard spheres*: a liquid of hard spheres with fluctuating, but additive, diameters.[83,84] At a rather low level, such a model might be appropriate for describing the influence of vibrating molecules on one another or the existence of a conformational equilibrium in the liquid. In any case, it has long been appreciated that the excess chemical potential of a mixture of hard spheres of different *fixed* diameters depends only on the first three moments of the diameter distribution, at least within the Percus–Yevick approximation.[85] But, from our comments in Section IV.A, this result tells us that the cavity distribution function [and thus $s(q)$] for *fluctuating* diameters, in this approximation, must also depend on these same three moments. Hence the equilibrium condition for a liquid of breathing hard spheres is the set of three equations

$$\langle q^n \rangle = \int dq \, s(q; \langle q \rangle, \langle q^2 \rangle, \langle q^3 \rangle) q^n,$$

$$n = 1, 2, 3.$$

Typically, the most physically revealing moments are not those arising from scalar degrees of freedom, but from vector ones. If $\vec{q}$ is a vector in an m-dimensional space, an obvious generalization of a second moment would be the $m \times m$ symmetric matrix

$$\mathbf{M}_{\alpha\beta} \equiv \langle q_{j(\alpha)} q_{j(\beta)} \rangle = \int d\vec{q} \, s(\vec{q}) q_{(\alpha)} q_{(\beta)},$$

$$\alpha, \beta = 1, \ldots, m, \tag{4.12}$$

which we could abbreviate in dyad notation

$$\mathbf{M} = \langle \vec{q}\vec{q} \rangle. \tag{4.13}$$

These kinds of matrices are interesting objects under both classical and quantal circumstances. Their symmetries and, more generally, the way in which elements of the form $\mathbf{M}_{\alpha\beta}$ depend on the α, β indices, carry a great deal of information about the behavior of the internal coordinate. In the fairly common situation that q is a vector in three dimensions, so that $\alpha, \beta = (x, y, z)$, and the liquid is isotropic, not only must M be symmetric, it must be true that

$$\mathbf{M}_{xx} = \mathbf{M}_{yy} = \mathbf{M}_{zz}, \qquad \mathbf{M}_{xy} = \mathbf{M}_{yz} = \mathbf{M}_{xz} = 0,$$

or in tensor form

$$\mathbf{M} = \tfrac{1}{3}\langle|\vec{q}|^2\rangle\mathbf{1}. \qquad (4.14)$$

However, if the liquid is not isotropic, that is if there is some *symmetry breaking* (as in the formation of a liquid crystalline phase), this matrix tells us what symmetry the system has. If, as in a nematic phase, there is a preferred axis in the system but not a preferred direction—if the system is aligned but not oriented—the matrix is still diagonal but the three axes are no longer equivalent

$$\mathbf{M}_{xx} \neq \mathbf{M}_{yy} \neq \mathbf{M}_{zz}, \qquad \mathbf{M}_{xy} = \mathbf{M}_{yz} = \mathbf{M}_{xz} = 0.$$

We could therefore arrive at a theory for the isotropic–nematic phase transitions by regarding the orientation of our molecules as the internal degree of freedom and the M matrix as the order parameter.

Another possibility of importance to us is that $\vec{q}$ is a vector of occupation numbers for some finite-state degree of freedom. In these cases the matrix M takes the form

$$\mathbf{M}_{\alpha\beta} = \langle v_{j\alpha}v_{j\beta}\rangle. \qquad (4.15)$$

It is easy to show that this M must also be diagonal. Since only one component of the $\vec{v}_j$ vector can be nonzero at a time, $v_{j\alpha}v_{j\beta} = 0$ unless $\alpha = \beta$. Furthermore, since $(v_{j\alpha})^2 = v_{j\alpha}$, the diagonal element is the occupation probability for that state

$$\mathbf{M}_{\alpha\beta} = \langle v_{j\alpha}\rangle\delta_{\alpha\beta}.$$

From Eq. (2.12) we also see that the trace of M must be 1.

When one graduates to quantum mechanical degrees of freedom in discretized path integral form, every internal coordinate is a vector coordinate q, even when the original (classical limit) variable is a scalar. Thus, the second moment matrix M is a now a symmetric $P \times P$ matrix[52]

$$\mathbf{M}_{t,t'} = \langle q_j^{(t)}q_j^{(t')}\rangle. \qquad (4.16)$$

The general form of the effective Hamiltonian for path integral problems (Eqs. 3.14–3.16), though, guarantees that $q^{(t)}$ components will have certain symmetries, regardless of the physical problem. In particular, the effective Hamiltonian is always invariant to translations in t, so that $\mathbf{M}_{t,t'} = \mathbf{M}_{t+T,t'+T}$. Combined with the fact that the order of the two components $q^{(t)}$ and $q^{(t')}$ is irrelevant, we see that the elements of the M matrix can only depend on the absolute value of the difference of the row and the column

$$M_{t,t'} = g(|t - t'|/P).$$ (4.17)

This symmetry is important for a number of reasons, one of which is that matrices of this special structure can always be diagonalized by Fourier transforms. Explicitly, any $P \times P$ matrix M satisfying Eq. (4.17) is converted into the $P \times P$ diagonal matrix of eigenvalues Λ

$$M = U^+\Lambda U,$$

by the $P \times P$ unitary matrix U

$$U_{t,t'} = P^{-1/2}e^{2\pi itt'/P}.$$ (4.18)

Because the underlying symmetry is a fundamental feature of the Hamiltonian it will frequently be the case that not only this moment, but all of the correlation functions and thermodynamic results can be diagonalized by these same matrices. Whenever that happens, we no longer have to treat a classical system with $N \times P$ internal degrees of freedom in order to handle the quantum mechanics of N atoms. *Instead our problem is broken up into P independent classical liquids, each with one internal degree of freedom per atom.* In effect we will have found the *normal modes* of the intramolecular variables in such cases.[5, 26-35, 52]

There is, of course, a physical reason for this behavior. As we noted in Section III, the t labels are a kind of time, so the property we are exploiting is a kind of time translation invariance. Real (physical) time has such a property when energy is conserved, but why should our artificial time behave this way? The answer is that the two kinds of time enter into the Feynman path integral formalism in precisely the same way: any quantum mechanical expression for time evolution at some constant energy, involving a real time t, has an equivalent equilibrium statistical mechanical expression relevant to a constant $\beta = (k_B T)^{-1}$ and an artificial time τ. This artificial time, which differs only by a factor of i from real, physical time is in fact the time that shows up in our formulas.[73]

The status of our time variable as an *imaginary time* can be seen quite simply by looking at the second moment matrix, Eq. (4.16). From the form for the partition function given in Eq. (3.10), it is easy to write our correlation function in operator language. All one has to do is to make use of the closure relation

$$\int dq_j^{(t')}\langle q_j^{(t)}|e^{-\beta\hat{H}/P}|q_j^{(t')}\rangle\langle q_j^{(t')}|e^{-\beta\hat{H}/P}|q_j^{(t'')}\rangle = \langle q^{(t)}|e^{-2\beta\hat{H}/P}|q^{(t'')}\rangle,$$

to integrate out all of the q components except t and t'

$$\langle q_j^{(t)}q_j^{(t')}\rangle = Q^{-1}\int dq_j^{(t)}\int dq_j^{(t')}\langle q_j^{(t)}|e^{-(\beta/P)(t'-t)\hat{H}}|q_j^{(t')}\rangle$$
$$\times q_j^{(t')}\langle q_j^{(t')}|e^{-(\beta/P)(P-t'+t)\hat{H}}|q_j^{(t)}\rangle q_j^{(t)}.$$

If we define $\tau \equiv (t' - t)/P$, the operator equivalent of our correlation function is

$$\langle q_j^{(t)}q_j^{(t')}\rangle = Q^{-1}Tr[e^{-\beta\hat{H}}e^{\beta\tau\hat{H}}q_je^{-\beta\tau\hat{H}}q_j]$$
$$= Q^{-1}Tr[e^{-\beta\hat{H}}q_je^{-\beta\tau\hat{H}}q_je^{\beta\tau\hat{H}}], \qquad (4.19)$$

which is precisely the quantum mechanical ensemble average

$$\langle q_jq_j(\tau)\rangle = \langle q_je^{-\beta\tau\hat{H}}q_je^{\beta\tau\hat{H}}\rangle. \qquad (4.20)$$

The significance of Eq. (4.20) for us is that it is virtually identical to the *real-time* autocorrelation function for a coordinate q_j

$$\langle q_jq_j(t)\rangle = \langle q_je^{it\hat{H}/\hbar}q_je^{-it\hat{H}/\hbar}\rangle.$$

Hence all of the insight into the *dynamics* of our system given by such a time correlation function is also given by the *equilibrium* second moment matrix $M_{t,t'} = g(\tau)$, with the identification

$$t = i(\hbar\beta)\tau. \qquad (4.21)$$

The matrix M itself is thus an imaginary-time correlation function that can be analytically continued to produce real-time information.

There are a variety of different ways to do this analytical continuation, depending on the basic computational scheme being employed for the rest of the problem. There are also different possible targets—in particular, one is likely to need frequency dependent information, such as the power spectrum

$$\hat{I}(\omega) = (2\pi)^{-1}\int_{-\infty}^{\infty} dt\, e^{-i\omega t}C(t).$$

rather than the time correlation function $C(t)$ itself. The optical absorption intensity, for example, is precisely $\hat{I}(\omega)$ if $C(t)$ is the dipole autocorrelation function.

One of the ways to calculate frequency-dependent responses is to look at integrals of the form

$$f(\omega_n) = \int_0^1 d\tau \, e^{-\hbar i \omega_n \beta \tau} C(\tau).$$

with $C(\tau)$ an imaginary-time correlation function. For suitable real frequencies ω_n (the Matsubaru frequencies), the $f(\omega_n)$ are the Fourier coefficients in the Fourier series representation of $C(\tau)$ over the interval $0 < \tau < 1$. By analytically continuing the Matsubaru frequencies

$$i\omega_n \rightarrow \omega \pm \varepsilon, \qquad \varepsilon \rightarrow 0^+,$$

one can calculate the power spectrum directly from the $f(\omega)$'s.[86,87] This approach was used successfully to study the optical absorption of liquids of polarizable molecules[26-34,88,89] (including collision-induced absorption)[35] and the spectra of solvated electrons.[90]

Alternatively, one can look at correlation functions in *complex* time.[91-93] An increasingly popular version of analytical continuation considers the correlation function

$$G(t) = Q^{-1} Tr[q e^{-\hat{H}T} q e^{-\hat{H}T^*}]$$

$$= Q^{-1} \int dq \int dq' \, qq' \, |\langle q | e^{-\hat{H}T} | q' \rangle|^2$$

$$T \equiv \tfrac{1}{2}\beta - (it/\hbar).$$

The power spectrum of $G(t)$ is trivially related to that of $C(t) = \langle qq(t) \rangle$ by

$$\hat{I}(\omega) = e^{\hbar \omega \beta / 2} \hat{G}(\omega),$$

but $G(t)$ is often easier to compute because it is more symmetric.[94,95] This fact has been especially helpful with reaction rate calculations.[96-98]

Interestingly, this symmetry shows up in the purely imaginary-time version[97]

$$G(\tau) = Q^{-1} Tr[q e^{-\beta((1/2)+\tau)\hat{H}} q e^{-\beta((1/2)-\tau)\hat{H}}],$$

which is formally related to $G(t)$ by the analytical continuation Eq. (4.21). Our previous imaginary-time correlation functions $C(\tau) = \langle qq(\tau) \rangle$ are invariant under the replacement $\tau \rightarrow 1 - \tau$, making them symmetric about $\tau = \tfrac{1}{2}$. The new function $G(\tau) = C(\tau \rightarrow \tfrac{1}{2} + \tau)$, however, is symmetric about $\tau = 0$. Thus, if we were to try to transform both imaginary-time functions back to real time with Eq. (4.21), we would find that $G(t)$ would have to be real, whereas $C(t)$ could be complex. [Parenthetically, Eq. (4.21) would seem to allow you to write $C(t)$ as $G(t \rightarrow t - i\tfrac{1}{2}\hbar\beta)$ and $G(t) = G(\tau \rightarrow t/(i\hbar\beta))$, so that we could analytically continue $C(\tau)$ simply by making substitutions.[99] The danger of this short cut is that one is not guaranteed to pick the correct branch of the answer.][90, 100]

In view of the success of analytical continuation, we should expect that the connections with spectroscopy afforded by these real-time versions of our correlation functions will usually be our most direct ties with experiment. However, we should also point out that the imaginary-time functions have their own physical meaning. In the classical limit, density matrices are completely diagonal—which implies that any two components of a q vector must be the same. As the system becomes more quantum mechanical, though, this correlation has to diminish, reaching a minimum level at the extreme quantum limit. Equivalently, if we adopt the argument given in Section III that the different t components represent quantal uncertainty, we see that maximum uncertainty would be portrayed by maximum diversity in the components— minimum correlation—whereas minimum uncertainty would demand total correlation. From this perspective, the M matrix is also telling us how quantum mechanical our internal degrees of freedom are.

Regardless of the interpretation, calculating what these functions look like in the condensed phase is a full many-body problem—an approach to which we discuss in the next section—but we can at least present a few examples of these imaginary-time corelation functions as they appear in the isolated atom. Without any interactions, the intramolecular distribution of a discretized coordinate q is

$$s_0(\boldsymbol{q}) = \rho_0(\boldsymbol{q}) \bigg/ \int d\boldsymbol{q} \, \rho_0(\boldsymbol{q}), \tag{4.22}$$

with ρ_0 given by Eq. (3.14). So, as we did above, we can evaluate our gas-phase correlation function by using the closure condition to integrate out most of the imaginary-time components. In terms of the gas-phase density matrices (Eqs. 3.12 and 3.13), we find

$$g_0(\tau) \equiv \langle g_j q_j(\tau) \rangle_0$$
$$= Q_0^{-1} \int dq_j^{(t)} \int dq_j^{(t')} q_j^{(t)} q_j^{(t')} \rho_0(q_j^{(t)}, q_j^{(t')}; \beta\tau) \rho_0(q_j^{(t)}, q_j^{(t')}; \beta(1 - \tau)) \tag{4.23}$$

with

$$Q_0 = Tr(e^{-\beta \hat{H}_{0j}}) = \int dq_j \, \rho_0(q_j, q_j; \beta), \tag{4.24}$$

the single-atom internal partition function.

Not surprisingly, this correlation function becomes especially simple when we have occupation numbers for our variables. Suppose, as usual, we have m possible internal states. Rather than work with the correlation function for any one state α,

$$g_{0\alpha}(\tau) \equiv \langle v_{j\alpha} v_{j\alpha}(\tau) \rangle_0, \tag{4.25}$$

it ends up being more useful to consider the symmetrized correlation function,

$$g_0(\tau) \equiv \sum_\alpha g_{0\alpha}(\tau) = \langle \vec{v}_j \cdot \vec{v}_j(\tau) \rangle_0 \tag{4.26}$$

or, equivalently,

$$C_0(\tau) \equiv [mg_0(\tau) - 1]/(m - 1). \tag{4.27}$$

To obtain these quantities all we have to do is to substitute the form for the density matrix discussed in Section III (with the j label understood)

$$\rho_0(\vec{v}, \vec{v}'; \beta\tau) = \sum_{\alpha, \beta} v_\alpha B_\tau(\alpha, \beta) v'_\beta,$$

$$B_\tau(\alpha, \beta) \equiv (e^{-\beta\tau H_0})_{\alpha, \beta},$$

into Eqs. (4.23) and (4.24) and make use of a few occupation number properties. The results are then just

$$g_0(\tau) = B_\tau(\alpha, \alpha) B_{1-\tau}(\alpha, \alpha)/B_1(\alpha, \alpha) \tag{4.28}$$

[which could also be written in terms of the a and c constants of Eq. (3.23) with the definition $B_\tau = \exp(a_\tau + c_\tau)$]. Nondiscrete variables, however, have to be dealt with on a case-by-case basis.

For example,

1. *Harmonic oscillator.* By directly integrating the density matrix presented in Section III one obtains

$$g_0(\tau) = \langle q^2 \rangle_0 \frac{\cosh[\frac{1}{2}x(1 - 2\tau)]}{\cosh \frac{1}{2}x},$$

$$\langle q^2 \rangle_0 = \frac{\hbar}{2m\omega_0} \tanh \frac{1}{2}x, \tag{4.29}$$

with a redefinition of the oscillator frequency as ω_0 and $x = \hbar\omega_0\beta$.

The *symmetrized* imaginary-time correlations function is thus

$$G_0(\tau) = \langle q^2 \rangle_0 \frac{\cosh x\tau}{\cosh \frac{1}{2}x},$$

which can be analytically continued to real time using Eq. (4.21) and then Fourier transformed to yield

$$\hat{G}_0(\omega) = \langle q^2 \rangle_0 \operatorname{sech} \frac{x}{2} [\delta(\omega - \omega_0) + \delta(\omega + \omega_0)].$$

Hence the power spectrum corresponding to $g_0(\tau)$ is

$$\hat{C}_0(\omega) = e^{\hbar\omega\beta/2} \hat{G}_0(\omega)$$

$$= \langle q^2 \rangle_0 \frac{1}{2} \left\{ \left(1 + \tanh \frac{x}{2} \right) \delta(\omega - \omega_0) + \left(1 - \tanh \frac{x}{2} \right) \delta(\omega + \omega_0) \right\} \quad (4.30)$$

2. *Two-state tunneling.*[52]   From Eqs. (3.23), (3.28), and (4.28), we get

$$g_0(\tau) = \cosh[\beta K\tau] \cosh[\beta K(1 - \tau)]/\cosh[\beta K],$$

$$C_0(\tau) = 2g_0(\tau) - 1 = \frac{\cosh \beta K(1 - 2\tau)}{\cosh \beta K}. \quad (4.31)$$

Clearly, if we set $2K = \hbar\omega_0$, the imaginary-time correlation function $C_0(\tau)$ for the two-state problem is identical to the $g_0(\tau)$ of the harmonic oscillator (except for the factor of $\langle q^2 \rangle_0$). The gas-phase power spectra of the two problems are therefore identical.[88,89,92]

3. *Electron correlation.*[5]   With Eq. (3.29) in place of Eq. (3.28), we get

$$g_0(\tau) = F[x\tau]F[x(1 - \tau)]/F[x],$$

$$C_0(\tau) = [4g_0(\tau) - 1]/3, \quad (4.32)$$

provided we redefine $x = \frac{1}{2}\beta\Delta E_{sp}$, and we say

$$F(y) \equiv \frac{1}{4}(e^y + 3e^{-y}).$$

We could use any of these examples as an illustration of our point that these correlation functions serve to monitor the "quantum mechanicality" of our system. Plotted in Fig. 3 is the correlation function $C_0(\tau)$ of Eq. (4.32) for

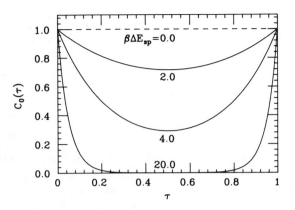

Figure 3. The imaginary-time correlation function for the electron-correlation model discussed in the text. The different curves are for different values of the gas-phase $s-p$ energy splitting, ΔE_{sp}, measured in units of $k_B T$. The dashed line is the classical limit $(\Delta E_{sp}/k_B T) = 0$.

different values of the reciprocal temperature parameter x. At $x = 0$, we have a completely classical situation, but as x increases, the energy spacing ΔE_{sp} goes up (relative to the $k_B T$), making the atom more quantal. Correspondingly, we see that the imaginary-time correlation function goes from being constant at 1 to being almost constant at 0 (except for $\tau = 0, 1$).

These features will, in fact be universal with the $g(\tau)$ and $C(\tau)$ occupation-number correlation functions appropriate to the full liquid as well. At the end points,

$$g_\alpha(\tau = 0, 1) = \langle v_{j\alpha}^2 \rangle = \langle v_{j\alpha} \rangle$$

so that Eq. (2.12) and the analogue of Eq. (4.27) imply

$$g(\tau = 0, 1) = 1,$$
$$C(\tau = 0, 1) = 1,$$
(4.33)

under all circumstances. However, in the *classical limit*, $v_j = v_j(\tau)$, which yields with the analogue of Eq. (4.26),

$$g(\tau) = \langle \vec{v}_j \cdot \vec{v}_j \rangle = 1,$$
$$C(\tau) = 1,$$

whereas, in the *extreme quantum mechanical limit* all the t components will be independent $(\tau \neq 0, 1)$:

$$\langle \vec{v}_j \cdot \vec{v}_j(\tau) \rangle = \langle \vec{v}_j \rangle \cdot \langle \vec{v}_j(\tau) \rangle.$$

If, in this limit, all m states are equivalent (as in our examples), $\langle v_{j\alpha} \rangle = 1/m$. Hence

$$g(\tau) = 1/m,$$
$$C(\tau) = 0.$$

This ability to study just how quantum mechanical our system is turns out to be rather enlightening, because the answer, surprisingly enough, is something that can change with the thermodynamic conditions.

V.   TECHNIQUES OF INTRAMOLECULAR STATISTICAL MECHANICS

In Section IV we identified the solvent correction to the intramolecular probability density, what we called the cavity distribution function $y(q)$, as the primary target in any study of an internal coordinate q. From $y(q)$ one can get the full liquid-state distribution $s(q)$ and hence all of the intramolecular moments and correlation functions. From Eq. (4.4), however, we see that the cavity distribution function can be written as the average Boltzmann factor for the interaction part of the Hamiltonian:

$$y(q_i) = \left\langle \exp\left\{ -\beta \sum_{k \neq i} v(r_{ik}; q_i, q_k) \right\} \right\rangle_{q_i}, \tag{5.1}$$

where the brackets refer to an average, at fixed q_i, over all the internal and translational coordinates of the remainder of the liquid, that is, an average subject to the Hamiltonian

$$H[q_i] \equiv \sum_{j \neq i} H_{0j}(q_j) + \sum_{j < k} u_0(r_{jk}) + \sum_{\substack{j < k \\ j, k \neq i}} v(r_{jk}; q_j, q_k). \tag{5.2}$$

The analogous form is even simpler for the infinitely dilute solute case, Eq. (3.1). Then there are no internal degrees of freedom in the solvent molecules, so that the cavity distribution for the solute internal coordinate is given by

$$y(q) = \langle \exp\{ -\beta V_I(q, R) \} \rangle_B,$$
$$= \left\langle \exp\left\{ -\beta \sum_{k \neq i} v(r_{ik}; q) \right\} \right\rangle_B, \tag{5.3}$$

with the average just an average over the bath Hamiltonian V_B given by Eq. (2.3) appropriately modified for a mixture)

$$V_B = \sum_{k \neq i} u_0^{uv}(r_{ik}) + \sum_{\substack{j < k \\ j, k \neq i}} u_0^{vv}(r_{jk}).$$ (5.4)

(Here u_0^{uv} and u_0^{vv} are the solute–solvent and solvent–solvent pair potentials, respectively.) As always, all of these classical–statistical–mechanical results continue to hold in quantum mechanical cases if we use discretized path integrals. Indeed, precisely the same formulas apply, but, as usual, the internal coordinates become vectors q, the interaction potentials v are transformed as shown in Eq. (3.16), and the internal Hamiltonian is $-k_B T \ln \rho_0(q)$.

This general problem of having to compute the average of a Boltzmann factor is an extremely common one in statistical mechanics. Depending on what kind of liquid theory we want to use, there are a number of ways to proceed. We shall start by detailing two.

A. Cumulant Expansion and Mean-Field Theory

It is straightforward to do a cumulant expansion of the average of any exponential

$$\langle \exp\{-\beta H\} \rangle = \exp\{-\beta \langle H \rangle + \tfrac{1}{2}\beta^2[\langle H^2 \rangle - \langle H \rangle^2] + \cdots\}.$$ (5.5)

Since the resulting terms in the exponent are ordered in the size of the fluctuations, a simple way to generate a mean-field theory is to truncate this series at the first term—ignoring the fluctuations.

Consider the single-solute case, Eq. (5.3). To lowest order in fluctuations,

$$y(q) = \exp\left\{-\beta \sum_{k \neq i} \langle v(r_{ik}; q) \rangle_B\right\},$$

which, by standard liquid-theory manipulations,[1,2] can be written

$$y(q) = \exp\left\{-\beta \rho \int d\vec{r}\, g_B^{uv}(r)v(r; q)\right\},$$ (5.6)

with $g_B^{uv}(r)$ the solute–solvent radial distribution function arising from Eq. (5.4) and ρ the solvent density. Remembering that cavity distribution functions are Boltzmann factors for excess chemical potentials, Eq. (4.5), we see that the excess chemical potential here is just what the average energy of interaction would be if the liquid structure were unperturbed by the interaction.

A similar result obtains when we have mutually interacting internal degrees of freedom. In mean-field theory,

$$y(q_i) = \exp\left\{-\beta\rho \int d\vec{r} \int dq\, s(q)g_{[q_i]}(r; q_i, q)v(r; q_i, q)\right\},$$

where $g_{[q_i]}$ is the radial distribution function relevant to Eq. (5.2) (defined as in Eq. 4.6). Such a result is still not all that tractable, but a reasonable approximation comes from realizing that the liquid structure around atom i, at least as produced by Eq. (5.2), is affected only indirectly by the presence of internal dynamics in the surroundings. It is therefore likely to be true that $g_{[q_i]} \approx g_B$, the radial distribution function from Eq. (2.3). Hence we can write the mean-field answer

$$y(q_i) = \exp\left\{-\beta\rho \int d\vec{r} \int dq\, s(q)g_B(r)v(r; q_i, q)\right\}. \tag{5.7}$$

Notice the appearance in this mutually interacting situation of $s(q)$, the internal distribution derived from the full Hamiltonian. The presence of this quantity in $y(q)$, which is itself an ingredient in $s(q)$, is a statement of the self-consistency central to these problems. In fact, the self-consistency is sufficiently straightforward in mean-field theory that it affords us an opportunity to recognize it as a rather common feature of many-body problems. As an illustration of how this self-consistency comes in, suppose we examine a classical version of the two-state tunneling model introduced previously: *a liquid of (classical) two-state molecules.*[38]

Rather than use occupation number notation, let us define the two states to be $\mu = \pm 1$. Then if the interaction is $-v(r)$ between atoms in like states and $+v(r)$ between atoms in unlike states, we have

$$v(r_{jk}; \mu_j, \mu_k) = -\mu_j\mu_k v(r),$$

so our model is really an Ising model in a liquid. According to our mean-field theory,

$$y(\mu_i) = \exp\left\{-\beta\rho \int d\vec{r} \sum_\mu s(\mu)g_B(r)v(r; \mu_i, \mu)\right\}$$

$$= \exp\{\mu_i\langle\mu\rangle\langle -2\beta\Delta U/N\rangle\},$$

where

$$\langle -\beta \Delta U/N \rangle = \beta \frac{1}{2}\rho \int d\vec{r}\, g_B(r)v(r)$$

is an average excess interaction energy. However, Eq. (4.3) implies that the moment $\langle \mu \rangle$ must satisfy the equation

$$\langle \mu \rangle = \sum_\mu s(\mu)\mu = \sum_\mu s_0(\mu)y(\mu)\mu \bigg/ \sum_\mu s_0(\mu)y(\mu).$$

If, for simplicity, we now limit ourselves to the case that the two states are degenerate in the gas phase, $s_0(\mu) = \frac{1}{2}$, then this self-consistency condition reads

$$\langle \mu \rangle = [y(+1) - y(-1)]/[y(+1) + y(-1)]$$
$$= \tanh[\langle \mu \rangle \langle -2\beta \Delta U/N \rangle].$$

This result should look extremely familiar to anyone who has ever worked with lattice-spin models. The lattice equivalent of the average energy $\langle -2\Delta U/N \rangle$ can be thought of as the interaction energy per "neighbor," J, multiplied by the number of neighboring sites, z, so that our formula

$$\langle \mu \rangle = \tanh[\langle \mu \rangle \beta Jz],$$

is nothing but the liquid equivalent of the well-known Curie–Weiss–Bragg–Williams mean-field theory for an Ising model.[101] Just as with lattice systems, our model should have a phase transition between a paramagnetic ($\langle \mu \rangle = 0$) phase and a ferromagnetic phase—so it should represent a ferrofluid (if the transition is not preempted by freezing of the liquid).[45,102]

As simple as mean-field theory is to implement, it often will not be sufficient to analyze our problems from this vantage point. Considerations of numerical accuracy aside, many problems of interest in liquid theory are fundamentally *fluctuation* phenomena and would therefore be totally missed by this approach. For example, polarizable nonpolar molecules have zero average dipole moment (an identically zero mean field), but still interact with each other because of correlations in their instantaneous dipoles. More generally, any time correlations are central, we should expect mean-field theory to be suspect.

One could, in principle, consider trying to remedy the problem by going to higher order in Eq. (5.5). The difficulty is that cumulants beyond the first bring in liquid correlation functions involving more than two particles. In the equivalent solid-state models, with near-neighbor interactions, there is no serious obstacle to pursuing this route. Indeed, it often seems to work rather well there.[59,103,104] But in liquids the absence of reliable information about three- and four-body correlations makes this strategy impractical (at least at present). We shall have to pursue another direction.

B. Charging

Another standard way of taking the average of an exponential is a technique known as parameter differentiation or "charging."[80] The idea is that an average of the form $\langle e^{-\beta V} \rangle$ taken with respect to a Hamiltonian H_0, can be written as the ratio of two partition functions differing in the value of a parameter λ

$$\langle e^{-\beta V} \rangle = Tr[e^{-\beta V} e^{-\beta H_0}]/Tr[e^{-\beta H_0}]$$

$$= Tr[e^{-\beta H(\lambda=1)}]/Tr[e^{-\beta H(\lambda=0)}]$$

$$\equiv Q(\lambda = 1)/Q(\lambda = 0),$$

where

$$H(\lambda) \equiv H_0 + \lambda V. \tag{5.8}$$

But since

$$\ln[Q(\lambda = 1)/Q(\lambda = 0)] = \int_0^1 d\lambda \, \partial \ln Q(\lambda)/\partial \lambda = \int_0^1 d\lambda \langle -\beta V \rangle_\lambda,$$

with the brackets defining an average with respect to $H(\lambda)$

$$\langle f \rangle_\lambda \equiv Tr[f e^{-\beta H(\lambda)}]/Tr[e^{-\beta H(\lambda)}], \tag{5.9}$$

our desired average of an exponential can always be written *rigorously* as the exponential of an average:

$$\langle e^{-\beta V} \rangle = \exp \left\{ \int_0^1 d\lambda \langle -\beta V \rangle_\lambda \right\}. \tag{5.10}$$

Furthermore, if the potential V is a sum of pair potentials, the average in the exponential will be a simple integral over a radial distribution function. Thus, the only problem is to find this radial distribution function for an arbitrary value of the parameter λ.

One can apply this result with any choice of λ (including the electric charge if there is one in the problem—hence the name), but to determine cavity distribution functions for internal degrees of freedom, the simplest procedure is to charge with respect to the tagged internal coordinate itself. Suppose we start with the definition of the cavity distribution function in Eq. (4.4):

$$y(q_1) = Q_{(-1)}^{-1} \int dq_{\neq 1} \int dR \exp\{-\beta H'(q, R)\}$$

$$= e^{-\beta \Delta \mu_{q1}}$$

with

$$H'(q, R) \equiv H(q, R) - H_{01}(q_1),$$

equal to the complete Hamiltonian (i.e., Eqs. 2.1–2.4 plus any necessary path-integral modifications) excluding the gas-phase Hamiltonian for q_1.

Following the same steps as we did above, we can write

$$\ln[y(q)/y(q^{ref})] = \int_{q_{ref}}^{q} dq_1 \, \partial \ln y(q_1)/\partial q_1$$

$$= \int_{q_{ref}}^{q} dq_1 \langle \partial(-\beta H')/\partial q_1 \rangle_{q_1},$$

where the brackets refer to an average, taken at constant q_1, with respect to the Hamiltonian H'. The derivative will only select out the pair interactions v between particle 1 and the other particles, so if we define

$$\phi(r_{jk}; q_j, q_k) \equiv -\beta v(r_{jk}; q_j, q_k) \tag{5.11}$$

and make use of the radial distribution function from Eq. (4.6), we get

$$y(q)/y(q^{ref}) = \exp\left\{\int_{q_{ref}}^{q} dq_1 \rho \int d\vec{r} \int dq_2 \, s(q_2) g(r; q_1, q_2) \frac{\partial \phi(r; q_1, q_2)}{\partial q_1}\right\}$$

$$= e^{-\beta(\Delta \mu_q - \Delta \mu_q^{ref})}. \tag{5.12}$$

Equation (5.12) is an exact result for any scalar internal degree of freedom. If the coordinate q is a vector coordinate, precisely the same expression can be derived if both the q_1 derivative and the q_1 integral are assumed to be with respect to the *magnitude* q_1 of the vector $\vec{q}_1$. That is, if we define

$$\vec{q}_1 \equiv q_1 \hat{q}_1,$$

and differentiate and integrate at fixed unit vector $\hat{q}_1$, we see that Eq. (5.12) holds for *any* classical or (path-integral-transformed) quantal internal degree of freedom in a neat liquid.

One immediate consequence of this finding is that we now know how to produce systematic corrections to mean-field theory. As before, we get a mean-field prediction by assuming that the structure of the liquid is unaffected by the internal dynamics:

$$g(r; q_1, q_2) \approx g_B(r), \tag{5.13}$$

with $g_B(r)$ the usual radial distribution function from Eq. (2.3). Substituting Eq. (5.13) into Eq. (5.12) and doing the q_1 integral gives us the same mean-field answer we arrived at earlier, Eq. (5.7). However, Eq. (5.13) is nothing but the first term in an exact *thermodynamic perturbation-theory* expansion of the full radial distribution function.[105] This expansion can be expressed as a sum of corrections ordered in powers of the ϕ functions from Eq. (5.11). In Section V.C we show how a portion of the these corrections can be summed exactly to infinite order.

Before doing so, though, it is worth presenting an example of the general structure one gets from this approach. Suppose we have a vector internal coordinate with an interaction of the form of Eq. (2.6). This form can, of course, represent a tensor interaction between classical three-dimensional vectors, but our discussion in the interim has shown that it also embodies *all* possible interactions between classical (and quantal) occupation-number vectors, as well as including some kinds of quantal interactions between continuous coordinates (Eqs. 2.13 and 3.18). Suppose further that the radial distribution function has a similar form:

$$g(r_{jk}; \vec{q}_j, \vec{q}_k) = g_B(r_{jk}) + \vec{q}_j \cdot G(r_{jk}) \cdot \vec{q}_k, \tag{5.14}$$

(which, in the case of classical occupation numbers will always be true). Then Eq. (5.12) tells us that

$$\Delta \mu_{\vec{q}} = \Delta \mu_{\vec{q}} \text{(mean-field)} + \int_0^q dq_1 \, \rho \int d\vec{r} \int d\vec{q}_2 \, s(\vec{q}_2)$$

$$(\hat{q}_1 \cdot T(r) \cdot \vec{q}_2)(\vec{q}_2 \cdot G(r) \cdot \vec{q}_1)$$

$$= \Delta \mu_{\vec{q}} \text{(mean-field)} + \frac{1}{2} \rho \int d\vec{r} \, \vec{q} \cdot T(r) \cdot \langle \vec{q}\vec{q} \rangle \cdot G(r) \cdot \vec{q},$$

with

$$\Delta \mu_{\vec{q}} \text{(mean-field)} = \vec{q} \cdot \rho \int d\vec{r} \, g_B(r) T(r) \cdot \langle \vec{q} \rangle$$

[Note that in the event that the internal coordinates are discrete, the integral over $\vec{q}_2$ has to be replaced by a sum, but the integral over the (magnitude) q_1 remains an integral.]

The basic structure of this answer is that the excess chemical potential is a sum of two terms

$$\Delta\mu_{\vec{q}} = \vec{q} \cdot \mathbf{H} + \vec{q} \cdot \mathbf{W} \cdot \vec{q}, \tag{5.15}$$

the first of which, the mean-field term, is proportional to $\langle \vec{q} \rangle$, and the second of which, a fluctuation term, is proportional to the $\mathbf{M} = \langle \vec{q}\vec{q} \rangle$ matrix introduced in Section IV. Both terms, moreover, are proportional to a part of the internal energy. This particular form has a number of interesting consequences. For one thing, if $\vec{q}$ really is a three-dimensional vector in an isotropic liquid, then the mean-field term will vanish (because $\langle \vec{q} \rangle = 0$) and the fluctuation term will reduce to a constant times q^2 (since $\mathbf{W}$ will have to be proportional to the unit tensor). This reduction is precisely what happens when $\vec{q}$ is a fluctuating dipole in treatments of polarizability[26-35] and electron correlation.[5]

On the other hand, if $\vec{q}$ is a path-integral vector $\mathbf{q}$, then we can take advantage of our ability to diagonalize path-integral expressions. Not only can the $\mathbf{M}$ matrices be diagonalized by the matrix $\mathbf{U}$ given by Eq. (4.18), but any path-integral ensemble average of the form $f(t, t')$—such as $\mathbf{W}$—will be diagonalized by the same matrix. The underlying imaginary-time translation symmetry of the effective Hamiltonian guarentees that. Hence, if we define the eigenvector

$$\mathbf{q} = \mathbf{U}^{+}\mathbf{q}, \tag{5.16}$$

and the diagonalized potential-of-mean-force matrix

$$W = \mathbf{U}^{+}\mathbf{W}\mathbf{U}, \tag{5.17}$$

the fluctuation term will be simply a sum over a contribution for each eigenmode

$$\Delta\mu_{\mathbf{q}} - \Delta\mu_{\mathbf{q}}\,(\text{mean-field}) = \sum_{s=1}^{P} W_{(s)}[\mathbf{q}^{(s)}]^2, \tag{5.18}$$

with each contribution a *classical average energy for that mode*

$$W_{(s)} = (\Lambda_{(s)}/P)\frac{1}{2}\rho \int d\vec{r}\, v(r)G_{(s)}(r). \tag{5.19}$$

Here $\Lambda_{(s)}$ and $G_{(s)}$ are the eigenvalues of the intramolecular correlation (M) and radial distribution function (G) matrices, respectively, and $v(r)$ is the interaction potential in Eq. (3.18).[5,52]

Admittedly, both the results for classical vector coordinates and those for quantal scalar coordinates depend on having a radial distribution function satisfying Eq. (5.14). Such $g(r)$'s, however, are a fairly common outcome from mean-spherical approximation calculations—as we shall now see.

C. Thermodynamic Perturbation Theory and the Mean-Spherical Approximation

In Section V.B we showed how the problem of calculating the intramolecular distribution is, in essence, that of calculating a radial distribution function for a liquid with internal degrees of freedom, $g(r_{jk}; q_j, q_k)$, and, in particular, that of calculating the change from the distribution $g_B(r)$. Computing such changes in liquid structure, though, are exercises in thermodynamic perturbation theory, so let us begin by reviewing a few pertinent features of that area.

In their classic review, Andersen et al.[105] explained how to write the exact radial distribution function $g(r)$, arising from pair potentials

$$u(r) = u_0(r) + v(r),$$

as a reference radial distribution function $g_0(r)$ [arising from $u_0(r)$ alone] plus a diagrammatic expansion ordered in powers of

$$\phi(r) = -\beta v(r) \tag{5.20}$$

bonds. One cannot usually do anything with the full expansion, but selected subsets can be summed to infinite order. Of special interest are the *optimized chain sum* diagrams, $C(r)$, included by summing all the chains of $\phi \times [\exp(-\beta u_0)]$ and/or $h_0 (\equiv g_0 - 1)$ bonds with all possible noncrossing internal decorations of $f_0 (\equiv \exp(-\beta u_0) - 1)$ bonds (subject to the additional provisos that there be at least one ϕ bond and that no vertex be intersected by *only* h_0 bonds). When u_0 is a hard-sphere potential and the reference radial distribution function is that calculated from Percus–Yevick theory,[1,2] the resulting radial distribution function

$$g(r) = g_0(r) + C(r) \tag{5.21}$$

is precisely that of the *mean-spherical approximation* (MSA).

This diagrammatic language is not the normal one employed in discussing the MSA.[1,2] More typically, the MSA is defined by the Ornstein–Zernike integral equation

$$h(r_{12}) = c(r_{12}) + \rho \int d\vec{r}_3 \, c(r_{13}) h(r_{23}), \qquad (5.22)$$

with the closure, for hard spheres of diameter σ

$$\begin{aligned} h(r) &= -1, & r < \sigma \\ c(r) &= \phi(r), & r > \sigma, \end{aligned} \qquad (5.23)$$

and with (as always) $h(r) \equiv g(r) - 1$. But, given that $h_0(r)$ and $c_0(r)$ are governed by the Percus–Yevick approximation for hard spheres

$$\begin{aligned} h_0(r_{12}) &= c_0(r_{12}) + \rho \int d\vec{r}_3 \, c_0(r_{13}) h_0(r_{23}), \\ h_0(r) &= -1, & r < \sigma, \\ c_0(r) &= 0, & r > \sigma, \end{aligned} \qquad (5.24)$$

and that

$$\begin{aligned} h(r) &= h_0(r) + C(r), \\ c(r) &= c_0(r) + \Phi(r), \end{aligned} \qquad (5.25)$$

the ordinary formulation of the MSA leads us directly to an integral equation which, Anderson et al.[105] were able to show, sums up the optimized chain diagrams. The diagrammatic MSA is thus exactly the same as the conventional one. When it happens that all the diagrams with h_0 bonds in the chain contribute zero (a frequent occurrence in intramolecular problems), this integral equation takes on the particularly simple form

$$\begin{aligned} C(r_{12}) &= \Phi(r_{12}) + \rho \int d\vec{r}_3 \, \Phi(r_{13}) C(r_{23}), \\ C(r) &= 0, & r < \sigma, \\ \Phi(r) &= \phi(r), & r > \sigma. \end{aligned} \qquad (5.26)$$

The extension of the diagrammatic version of the MSA to internal-degree-of-freedom problems is quite easy. The Chandler–Pratt intramolecular statistical mechanics theorem[79] says that the diagrams need only be augmented by redefining the vertices as shown in Eq. (4.9) and taking the ϕ-bonds to be those of Eq. (5.11). However, the only places where the intramolecular coordinates enter are in the chains of ϕ and h_0 bonds. Thus, the only contributions of the internal dynamics to the optimized chain sum will be due to vertices at which two ϕ bonds intersect, leading to a contribution of

$$\Gamma_2 = \int dq_j \, s(q_j)\rho \int d\vec{r}_j \, \phi(r_{ij}; q_i, q_j)\phi(r_{jk}; q_j, q_k),$$

and due to the vertices at which a ϕ bond intersects an h_0 bond, giving

$$\Gamma_1 = \int dq_j \, s(q_j)\rho \int d\vec{r}_j \, \phi(r_{ij}; q_i, q_j)h_0(r_{jk}).$$

Under the circumstances discussed in Section V.B that our interaction is of the form of Eq. (2.6), a striking simplification takes place in these vertices; all of the internal coordinate contributions factor into moments

$$\Gamma_1 = \rho \int d\vec{r}_j \, \vec{q}_i \cdot \phi(r_{ij}) \cdot \langle \vec{q} \rangle h_0(r_{jk}),$$

$$\Gamma_2 = \rho \int d\vec{r}_j \, \vec{q}_i \cdot \phi(r_{ij}) \cdot \langle \vec{q}\vec{q} \rangle \cdot \phi(r_{jk}) \cdot \vec{q}_k,$$

with $\phi(r) = -\beta T(r)$. Once again, if $\vec{q}$ is a three-dimensional vector in an isotropic liquid, the moments themselves are simple

$$\langle \vec{q} \rangle = 0,$$

$$\langle \vec{q}\vec{q} \rangle = \langle q^2 \rangle \langle \hat{\Omega}\hat{\Omega} \rangle,$$

where the last average, over unit orientational vectors $\hat{\Omega}$, is proportional to the unit matrix

$$\langle \hat{\Omega}\hat{\Omega} \rangle = \int d\hat{\Omega}/4\pi \, \hat{\Omega}\hat{\Omega} = \frac{1}{3}\mathbf{1}.$$

So, we can write

$$\Gamma_1 = 0,$$

$$\Gamma_2 = \langle q^2 \rangle (q_i q_k)\rho \int d\vec{r}_j \int d\hat{\Omega}_j/4\pi \, \phi(r_{ij}, \hat{\Omega}_i, \hat{\Omega}_j)\phi(r_{jk}; \hat{\Omega}_j, \hat{\Omega}_k),$$

with $\phi(r; \hat{\Omega}_j, \hat{\Omega}_k) = -\beta\hat{\Omega}_j \cdot T(r) \cdot \hat{\Omega}_k$, a ϕ bond for atoms *without* their original internal degrees of freedom, but with artificial orientational unit vectors $\hat{\Omega}$ assigned to them.[24,25]

If we put these vertices back in their rightful places in the diagrams, we see that the optimized chain sum in this case is thus just a sum of chains of $\phi(r_{jk}; \hat{\Omega}_j, \hat{\Omega}_k)$ bonds, with a factor of $\langle q^2 \rangle$ for every black circle in the diagrams. Equivalently, we may absorb this factor into the ϕ bonds by assigning a magnitude of $\bar{q} \equiv \langle q^2 \rangle^{1/2}$ to each unit orientational vector in each bond. *The end result is that each nonvanishing term in the optimized chain sum, and therefore the sum as a whole, is what the chain sum would have been with the fluctuating internal coordinate $\vec{q}$ replaced by a coordinate fixed in magnitude at $\bar{q}$.* In other words, the final result for the radial distribution function is of precisely the form of Eq. (5.14),

$$g(r_{jk}; \vec{q}_j, \vec{q}_k) = g_B(r_{jk}) + C(r_{jk}; \vec{q}_j, \vec{q}_k)$$

$$C(r_{jk}; \vec{q}_j, \vec{q}_k) = \vec{q}_j \cdot G(r_{jk}) \cdot \vec{q}_k$$

with G defined so that

$$C_{\bar{q}}(r_{jk}; \hat{\Omega}_j, \hat{\Omega}_k) \equiv \bar{q}\hat{\Omega}_j \cdot G(r_{jk}) \cdot \bar{q}\hat{\Omega}_k$$

is the optimized chain sum (with no h_0 bonds) for an entire liquid of atoms without fluctuating internal degrees of freedom. In place of the fluctuating $\vec{q}$ vectors, there are freely rotating vectors of length $\bar{q}$.

What this development accomplishes is to convert the problem of calculating the radial distribution function for a liquid with internal dynamics into one without such dynamics. The remaining task of computing the fixed $\bar{q}$ chain sum is an ordinary MSA-like problem,[106] solvable via the integral equation (5.26), or through its angle-dependent version

$$C_{\bar{q}}(r_{12}; \hat{\Omega}_1, \hat{\Omega}_2) = \Phi(r_{12}; \hat{\Omega}_1, \hat{\Omega}_2) + \rho \int d\vec{r}_3 \int d\hat{\Omega}_3/4\pi \, \Phi(r_{13}; \hat{\Omega}_1, \hat{\Omega}_3)$$

$$C_{\bar{q}}(r_{23}; \hat{\Omega}_2, \hat{\Omega}_3),$$

$$C_{\bar{q}}(r; \hat{\Omega}_1, \hat{\Omega}_2) = 0, \qquad\qquad r < \sigma,$$

$$\Phi(r; \hat{\Omega}_1, \hat{\Omega}_2) = \bar{q}^2 \phi(r; \hat{\Omega}_1, \hat{\Omega}_2), \qquad r > \sigma.$$

Virtually these same manipulations can be repeated, with much the same profit, if one is interested in path-integral transformed quantum mechanical coordinates rather than classical vectors.[5,52] Again assuming an interaction of the Eq. (3.18) structure, one now gets a factor of the imaginary-time correlation matrix M at every vertex joining two ϕ bonds in our diagrams. Therefore, a diagram consisting of a chain of $(n + 1)\phi(r; q_1, q_2)$ bonds will be equal to the same diagram *without* internal coordinates (with the ϕ bonds prescribed by Eq. 5.20), but with an overall factor of

$$q_j \cdot M^n \cdot q_k \, P^{-(n+1)}.$$

Diagonalizing this factor as we did in Section V.B, though, makes it into a sum over modes

$$\sum_{s=1}^{P} q_j^{(s)} [\Lambda_{(s)}]^n q_k^{(s)} P^{-(n+1)}.$$

Now consider these modes one at a time. Since we can absorb a factor of $[\Lambda_{(s)}/P]$ into each ϕ bond, all the diagrams corresponding to the sth mode are just

$$q_j^{(s)} q_k^{(s)} \Lambda_{(s)}^{-1}$$

times what they would have been without internal degrees of freedom—but with an effective ϕ bond

$$\phi_{eff}^{(s)}(r) = [\Lambda_{(s)}/P]\phi(r). \tag{5.27}$$

Thus, for a quantum mechanical internal degree of freedom, the sum of all the diagrams (having no h_0 bonds) in the optimized chain sum will be a sum of optimized chain sums for effective classical systems without any internal degrees of freedom, one classical chain sum for each path-integral mode:

$$C(r_{jk}; q_j, q_k) = \sum_{s=1}^{P} q_j^{(s)} q_k^{(s)} C_{(s)}(r_{jk})/\Lambda_{(s)}. \tag{5.28}$$

The notation $C_{(s)}$ here refers to the classical chain sum which solves the integral equation (5.26) with ϕ taken to be $\phi_{eff}^{(s)}$. However, Eq. (5.28) also implies that our radial distribution function is of the standard form, Eq. (5.14), with the eigenvalues of the G matrix given by

$$G_{(s)}(r) = C_{(s)}(r)/\Lambda_{(s)}.$$

Hence we are in a position to evaluate our results from the last section for the *excess chemical potential*. Substituting into Eq. (5.19), we find that each mode leads to a factor

$$W_{(s)} = \Lambda_{(s)}^{-1} \frac{1}{2} \rho \int d\vec{r}\, \phi_{eff}^{(s)}(r) C_{(s)}(r)$$

$$= \Lambda_{(s)}^{-1} \langle -\beta \Delta U/N \rangle_{\phi_{eff}^{(s)}}. \tag{5.29}$$

What this expression tells us immediately is that each mode separately contributes an amount proportional to the excess internal energy of a simple classical liquid: for mode s, the excess is that of a liquid with pair potentials

$$u(r) = u_0(r) + (\Lambda_{(s)}/P)v(r),$$

over that of a liquid with pair potentials $u_0(r)$.[106] However, it also shows us how the self-consistency characteristic of intramolecular statistical mechanics enters a quantum problem. The imaginary-time correlation function eigenvalues $\Lambda_{(s)}$ are an essential part of the excess chemical potential, but they must be determined from the second moment matrix M, which itself comes from $s(q)$ and thus the excess chemical potential.

In effect, both of these results just flesh out our discussion of path-integral modes in Section IV. We pointed out there that the *existence* of these modes will always be fundamental to quantal problems. What we are adding here is a statement of precisely *how* the modes can come into play. It should probably be pointed out in this connection that it was the simple quadratic form of the potential of mean force, Eq. (5.15), that allowed us to achieve the required diagonalization so easily. This form, though, is a direct consequence of the MSA liquid theory, so let us close this section with a few remarks about the nature of the MSA.

The principal advantages of the MSA have always been the availability of simple analytical formulas for interesting model problems. The price that is paid for this simplicity is that one is confined to what is, in some sense, a linear treatment. From a diagrammatic perspective, the only direct interactions between particles that the MSA looks at are those linear in ϕ (whereas an exact theory should have $f = \exp(\phi) - 1$ functions, and therefore bonds representing *all* the powers of ϕ). It is true that the indirect terms involving other particles do give contributions to all orders, but the treatment of the direct terms is crucial to the methodology of this section. Without the limitation to precisely two ϕ bonds intersecting at a vertex, we would no longer be able to compress all of the intramolecular behavior into 2nd moments, at least not in general (occupation number variables are a conspicuous exception).

Perhaps the way to think about the situation is to regard the MSA, being a linear theory, as a small-oscillation theory for harmonic path integral modes. Applying some of the better theories currently in use for liquids without internal degrees of freedom should eventually help us to put in the anharmonic corrections.[25]

VI. GENERAL FEATURES OF INTRAMOLECULAR
BEHAVIOR IN LIQUIDS

Over the last decade the statistical mechanical approaches we have been discussing have been brought to bear repeatedly on a rather limited set of simple models. The aim in each case was not to achieve an ultimate theory, but to gather the first bits of insight into intramolecular phenomena in liquids. Ten years ago there were quite a few unanswered questions about liquids with internal degrees of freedom, and the questions were surprisingly basic ones: Should the liquid behave any differently because its atoms have internal dynamics? How does something as classical as an ordinary liquid interact with a quantum mechanical internal degree of freedom? The studies of the simple models provided the first microscopic routes to answering these questions.

It is somewhat presumptuous to try to read too much into the results from these few examples, but a number of results have shown up repeatedly and unambiguously over the course of these studies. Emboldened by this observation, we have collected some of the themes that seem to be essential to at least the cases examined so far. All of these generalizations should not be read as universal theorems, but they do form a set of useful concepts against which we can try to match new problems as they come along. Several of them in particular make specific testable predictions.

1. *Liquids with mutually interacting internal degrees of freedom are fundamentally self-consistency problems.* It stands to reason that if the solvent surrounding any given fluctuating molecule is composed of identical fluctuating molecules, any probability density for the tagged molecule will depend on the analogous probability density for the solvent molecules. That is, it will depend on itself. In the Chandler and Pratt language,[79] this idea is expressed by saying that the liquid-phase, intramolecular distribution of an internal coordinate q is given by

$$s(q) \propto s_0(q) y_{[s(q)]}(q),$$

where $y(q)$, the solvent-induced correction from the gas-phase distribution $s_0(q)$, is explicitly written as a functional of $s(q)$. More physically, we might say that the internal coordinates act to influence one another in a highly cooperative fashion.

Some of what makes this result nontrivial we have already discussed, but two other features are worth emphasizing. One is that the result is *exact*. The analogy we made with the mean-field theory of the Ising model and maybe even our continued use of the mean-spherical approximation are somewhat misleading in this regard. It is true that both of these involve simple-self-consistency relations for moments of the intramolecular distribution. It is also true that a self-consistency condition involving the magnetization (a first moment) is a hallmark of mean-field theories for spin systems. Nonetheless self-consistency of $s(q)$ is a rigorous requirement. We have simply been fortunate that what would, in general, be a functional equation of the form $s(q) = F[s(q)]$ ended up being an equation for moments of $s(q)$. Of course, for occupation-number variables, one never gets anything more complicated than moment relations.

A second point that we should not lose track of is that this result holds equally well for quantum mechanical systems—at least when they are transformed into a path-integral form. If we were to look at the equivalent formulas in operator notation, and to pay the proper respect to the noncommutativity of the operators (which, after all, is the origin of the quantal character), it might not have been so clear how to formulate such a self-consistency. Nor, perhaps, would it have been obvious that the full range of behaviors, from coherently delocalized excitations to purely localized, independent degrees of freedom, would have to be included within this same framework.

2. *Liquids with internal degrees of freedom are equivalent to (effective) liquids without internal degrees of freedom.* The earliest statement of this point was that a liquid composed of molecules capable of fluctuating between several possible states was fundamentally equivalent to a mixture containing fixed mole fractions of the molecules in those same states. As we did above, we should point out that this mixture–chemical equilibrium relationship also holds in quantum mechanical situations. More than that, though, we should also mention that this same basic point can show up in some fairly unexpected ways within a given theory.

In the MSA, certain kinds of interactions allow us to completely remove the fluctuating internal coordinates from the calculation of the radial distribution function. The role of the dynamical variable is then played by some moment of the internal distribution. In such cases our liquid is equivalent to a liquid with an effective interaction, or an effective density. For example, the properties of the solution of interconverting Cu^+ and Cu^{+2} ions that we discussed in Section II would be formally equivalent to one featuring Cu^{+x} ions $(1 < x < 2)$.[38]

3. *Mutually interacting quantum mechanical internal degrees of freedom can be diagonalized into normal modes. Each mode corresponds to an (effective) liquid with fluctuating classical internal degrees of freedom.* The underlying symmetry displayed by path integrals under translation in imaginary time *always* makes it possible to define a set of independent normal modes. There is no necessity for the single-molecule Hamiltonian or the interaction to be quadratic, nor need we adopt any approximate theories. However, in the MSA it is particularly easy to find these modes and, once they are found, their role is

clear. The diagonalization into MSA modes expresses the thermodynamics of the quantum liquid as a sum of contributions from P-independent classical liquids. Moreover, the MSA liquid problem for each classical mode can be turned into an effective liquid problem with no internal dynamics at all (point 2 above).

One can actually take these ideas somewhat further. As the classical limit is approached, the information contained in the P different liquids must become redundant—since it only takes a single classical liquid to describe a classical liquid. One can show, in fact, that there is always a single mode that evolves into the classical limit: The imaginary-time symmetry dictates that the particular vector of length P

$$\mathbf{q} = P^{-1/2}(1, 1, \ldots, 1)$$

is always a (normalized) eigenvector, and that its scaled eigenvalue, $\Lambda_{(s)}/P$, is always

$$\chi = P^{-1} \sum_{t'=1}^{P} \mathbf{M}_{t,t'} = \int_0^1 C(\tau) \, d\tau, \tag{6.1}$$

(the largest eigenvalue of the $\mathbf{M} = \langle \mathbf{qq} \rangle$ matrix). Since imaginary-time correlations become perfect in the classical limit, this χ must approach the classical answer of $\langle q^2 \rangle$ in that limit (which can be shown to imply that the other eigenvalues must go to zero). Thus this mode is the one that becomes classical when the system does.[52]

4. *There is a competition between quantum effects and condensed-phase effects.* This rather brash statement holds whenever the Hamiltonian for a system with internal degrees of freedom can be written in the form used in Section II. That is, it is true if it is possible to find a basis in which only the intramolecular parts are off diagonal, while the interactions remain diagonal. We should note that there are physical situations that do not permit the choice of such a basis. In particular, a tight-binding model for band structure cannot be cast into this form. However, a good many problems do fit this mold, and for good physical reasons.

Consider problems in which the internal coordinate interacts with its environment mostly via electrostatics—say, the internal dynamics leads to fluctuating charges. It is a simple result from electrostatics that the greater the concentration of charge, the stronger the interaction. Yet, what quantum mechanics does is to delocalize charge. In a general sort of way, one can see that when the uncertainty principle causes the coordinate q to be spread over space, instead of confined to a point, it is delocalizing whatever charges or multipole moments are associated with it. One can be even more precise within the path integral formalism. The width of the distribution of $q^{(t)}$ components is a measure of this spread.

Once we let quantum effects weaken the intermolecular interactions, the competition idea follows as a simple corollary. If quantum mechanics makes an internal coordinate interact more weakly with the rest of the liquid, it has to diminish the magnitude of the solvent effects. Conversely, since (all other things being equal) the liquid will try to get the most favorable electrostatic energy, it follows that the system can lower its energy by concentrating the charge—decreasing the spread in the $q^{(t)}$ components. Hence the liquid will make a quantal internal degree of freedom less quantal.

Path integrals allow us to view this competition in several ways. In a more traditional quantum mechanics language, one might have said that the conflict is between the desire to lower the kinetic energy (which the uncertainty principle says will be the outcome of delocalizing a coordinate) and the desire to lower the potential energy. With discretized path integrals, though, we see that delocalizing q is equivalent to increasing the entropy associated with the $q^{(t)}$ components, which makes the *free energy* of the purely classical path-integral system the final arbiter.[3]

This particular path-integral viewpoint allows us to test our ideas in a fairly specific way. If it is really true that a solvent will make an internal coordinate less quantum mechanical, an unavoidable implication is that a solvent has to increase the correlation between imaginary-time components. One should therefore always be able to look at correlation functions such as $C(\tau)$, or susceptibilities such as the χ defined in Eq. (6.1), and see them increase with solvent density. This increase is in fact seen in all the examples studied to date. The electron correlation studies[5] show that a bound electron, given a choice between being delocalized in an orbital without a dipole or somewhat more localized in a hybrid orbital with a dipole, opts more and more for the latter as the density increases. Similarly, the work on tunneling[52] finds that solvation makes a molecule prefer less and less to be in a symmetric combination of double-well states when it can localize itself in just one of the states and create a dipole. In both cases, imaginary-time correlations provided the evidence that these physical ideas were correct.

We should note, in closing, that the restriction on applying these concepts to inappropriate Hamiltonians is a genuine one. Hamiltonians whose excitations are macroscopically extended, such as the aforementioned tight-binding models or quantal Heisenberg models (with spin-wave excitations) fall outside the scope of this generalization. Nonetheless, we should also point out that the validity of the overall notion of competition is by no means limited to examples involving electrostatics. It even holds for interactions well described by hard cores—including the exchange forces which make atoms and electrons keep their distances from the centers of other atoms. With such excluded volume interactions, the idea is that a delocalized particle (e.g., an electron) occupies more space than the corresponding localized particle would. Hence, the need to maximize the entropy of the solvent (the packing forces) will also lead to some localization of the particle.[107]

Acknowledgments

The papers which first described the work covered in this review acknowledged the assistance of a number of people. I thank them again for their insights and comments. However, I am especially pleased to acknowledge the contributions of my collaborators on these and closely related projects: Professor Steven Desjardins, Dr. Vladimir Dobrosavljevic, Dr. Bing-Chang Xu, Steven Simon, and Charles Henebry. I also thank Professors Robert Harris and David Logan for their continued interest. This work was supported by NSF grant No. CHE-8815163.

References

1. J. P. Hansen and I. R. McDonald, *Theory of Simple Liquids*, 2nd ed., Academic, London, 1986.
2. H. L. Friedman, *A Course in Statistical Mechanics*, Prentice-Hall, Englewood Cliffs, NJ, 1985.
3. D. Chandler and P. G. Wolynes, *J. Chem. Phys.* **74**, 4078 (1981).
4. M. Herman and B. J. Berne, *Chem. Phys. Lett.* **77**, 163 (1981).
5. B.-C. Xu and R. M. Stratt, *J. Chem. Phys.* **89**, 7388 (1988).
6. W. L. Jorgenson, *J. Phys. Chem.* **87**, 5304 (1983)
7. D. A. Zichi and P. J. Rossky, *J. Chem. Phys.* **84**, 1712 (1986).
8. D. W. Oxtoby, *Adv. Chem. Phys.* **40**, 1 (1979).
9. D. W. Oxtoby, *Ann. Rev. Phys. Chem.* **32**, 77 (1981).
10. D. W. Oxtoby, *Adv. Chem. Phys.* **47**, 487 (1981),
11. K. S. Schweizer and D. Chandler, *J. Chem. Phys.* **76**, 2296 (1982).
12. L. R. Pratt and D. Chandler, *J. Chem. Phys.* **72**, 4045 (1980).
13. P. H. Berens and K. R. Wilson, *J. Chem. Phys.* **74**, 4872 (1981).
14. K. S. Schweizer, *Chem. Phys. Lett.* **125**, 118 (1986).
15. V. Dobrosavljevic and R. M. Stratt, *Phys. Rev. B* **35**, 2781 (1987).
16. V. Dobrosavljevic, C. W. Henebry, and R. M. Stratt, *J. Chem Phys.* **88**, 5781 (1988).
17. B. Raz and J. Jortner, *Proc. R. Soc. London Ser. A* **317**, 113 (1970).
18. J. Jortner and A. Gaathon, *Can. J. Chem.* **55**, 1801 (1977).
19. I. Messing, B. Raz, and J. Jortner, *Chem. Phys.* **25**, 55 (1977).
20. R. A. Chiles, G. A. Jongeward, M. A. Bolton, and P. G. Wolynes, *J. Chem. Phys.* **81**, 2039 (1984).
21. R. W. Hall and P. G. Wolynes, *J. Chem. Phys.* **83**, 3214 (1985).
22. M. Sprik, R. W. Impey, and M. L. Klein, *Phys. Rev. Lett.* **56**, 2326 (1986).
23. V. Dobrosavljevic, C. W. Henebry, and R. M. Stratt, *J. Chem. Phys.* **91**, 2470 (1989).
24. L. R. Pratt, *Mol. Phys.* **40**, 347 (1980).
25. J. S. Hoye and G. Stell, *J. Chem. Phys.* **73**, 461 (1980).
26. J. S. Hoye and G. Stell, *J. Chem. Phys.* **75**, 5133 (1981).
27. J. S. Hoye and K. Olaussen, *J. Chem. Phys.* **77**, 2583 (1982).
28. M. J. Thompson, K. S. Schweizer, and D. Chandler, *J. Chem. Phys.* **76**, 1128 (1982).
29. D. Chandler, K. S. Schweizer, and P. G. Wolynes, *Phys. Rev. Lett.* **49**, 1100 (1982).
30. K. S. Schweizer and D. Chandler, *J. Chem. Phys.* **78**, 4118 (1983).
31. K. S. Schweizer, *J. Chem. Phys.* **85**, 4638 (1986).
32. D. E. Logan, *Mol. Phys.* **46**, 1155 (1982).

33. D. E. Logan, *Mol. Phys.* **51**, 1365 (1984).
34. D. E. Logen, *Mol. Phys.* **51**, 1395 (1984).
35. D. E. Logan, *Chem. Phys. Lett.* **112**, 335 (1984).
36. M. Sprik and M. L. Klein, *J. Chem. Phys.* **89**, 7556 (1988).
37. S. Baer, *Mol. Phys.* **65**, 263 (1988).
38. R. M. Stratt, *J. Chem. Phys.* **80**, 5764 (1984).
39. F. Brouers, Ch. Holzhey, and J. Franz, in *Excitations in Disordered Systems*, M. F. Thorpe (Ed.), Plenum, New York, 1982.
40. L. A. Turkevich and M. H. Cohen, *Phys. Rev. Lett.* **53**, 2323 (1984).
41. L. A. Turkevich and M. H. Cohen, *Ber. Bunsenges. Phys. Chem.* **88**, 292 (1984).
42. D. E. Logan and P. P. Edwards, *Phil. Mag. B* **53**, L23 (1986).
43. R. W. Hall and P. G. Wolynes, *Phys. Rev. B* **33**, 7879 (1986).
44. T. V. Ramakrishnan, in *The Metallic and Nonmetallic States of Matter*, P. P. Edwards and C. N. R. Rao (Eds.), Taylor and Francis, London, 1985.
45. J. S. Hoye and G. Stell, *Phys. Rev. Lett.* **36**, 1569 (1976).
46. E. Martina and G. Stell, *J. Stat. Phys.* **27**, 407 (1982).
47. S. L. Carnie and G. Stell, *Phys. Rev. B* **26**, 1389 (1982).
48. N. E. Frankel and C. J. Thompson, *J. Phys. C* **8**, 3194 (1975).
49. P. C. Hemmer and D. Imbro, *Phys. Rev. A* **16**, 380 (1977).
50. L. Feijoo, C.-W. Woo, and V. T. Rajan, *Phys. Rev. B* **22**, 2404 (1980).
51. R. M. Stratt, *Phys. Rev. Lett.* **53**, 1305 (1984).
52. S. G. Desjardins and R. M. Stratt, *J. Chem. Phys.* **81**, 6232 (1984).
53. P. Ballone, Ph. de Smedt, J. L. Lebowitz, J. Talbot, and E. Waisman, *Phys. Rev. A* **35**, 942 (1987).
54. P. de Smedt, P. Nielaba, J. L. Lebowitz, J. Talbot, and L. Dooms, *Phys. Rev. A* **38**, 1381 (1988).
55. A. D. J. Haymet, M. R. Kramer, and C. Marshall, *J. Chem. Phys.* **88**, 342 (1988).
56. J. Juanos i Timoneda and A. D. J. Haymet, *J. Chem. Phys.* **90**, 1901 (1989).
57. F. Y. Wu, *Rev. Mod. Phys.* **54**, 235 (1982).
58. P. G. de Gennes, *Solid State Commun.* **1**, 132 (1963).
59. R. M. Stratt, *J. Chem. Phys.* **84**, 2315 (1986).
60. R. M. Stratt and B.-C. Xu, *Phys. Rev. Lett.* **62**, 1675 (1989).
61. B.-C. Xu and R. M. Stratt, *J. Chem. Phys.* **91**, 5613 (1989).
62. E. N. Economou, *Green's Functions in Quantum Physics*, Springer-Verlag, Berlin, 1983, pp. 7 and 8 and Chapter 5.
63. W. Klein, *Phys. Rev. B* **26**, 2677 (1982).
64. W. Klein and G. Stell, *Phys. Rev. B* **32**, 7538 (1985).
65. J. A. Given and W. Klein, *J. Chem. Phys.* **90**, 1116 (1989).
66. J. A. Given, *J. Chem. Phys.* **90**, 5068 (1989).
67. B. I. Balberg, *Phil. Mag. B* **56**, 991 (1987).
68. G. Stell, *J. Phys. A* **17**, L855 (1984).
69. R. M. Stratt, *J. Chem. Phys.* **67**, 5894 (1977).
70. D. Thirumalai, R. W. Hall, and B. J. Berne, *J. Chem. Phys.* **81**, 2523 (1984).
71. J. D. McCoy, S. W. Rick, and A. D. J. Haymet, *J. Chem. Phys.* **90**, 4622 (1989).
72. J. C. Tully, in *Dynamics of Molecular Collisions*, Part B, W. H. Miller (Ed.), Plenum, New York, 1976.
73. R. P. Feynman and A. R. Hibbs, *Quantum Mechanics and Path Integrals*, McGraw-Hill, New York, 1965, Chapter 10.
74. R. P. Feynman, *Statistical Mechanics*, W. A. Benjamin, Reading, MA, 1972, Chapters 2 and 3.
75. K. S. Schweizer, R. M. Stratt, D. Chandler, and P. G. Wolynes, *J. Chem. Phys.* **75**, 1347 (1981).

76. B. J. Berne and D. Thirumalai, *Ann. Rev. Phys. Chem.* **37**, 401 (1986).
77. H. F. Trotter, *Proc. Am. Math. Soc.* **10**, 545 (1959).
78. L. S. Schulman, *Techniques and Applications of Path Integration*, Wiley, New York, 1981.
79. D. Chandler and L. R. Pratt, *J. Chem. Phys.* **65**, 2925 (1976).
80. D. Chandler, in *The Liquid State of Matter: Fluids Simple and Complex*, E. W. Montroll and J. L. Lebowitz (Eds.), North Holland, Amsterdam, 1982.
81. L. Onsager, *Ann. NY Acad. Sci.* **51**, 627 (1949).
82. L. R. Pratt and D. Chandler, *J. Chem. Phys.* **66**, 147 (1977).
83. S. G. Desjardins and R. M. Stratt, *J. Chem. Phys.* **90**, 6809 (1989).
84. D. A. Kofke and E. D. Glandt, *J. Chem. Phys.* **90**, 439 (1989).
85. J. L. Lebowitz and J. S. Rowlinson, *J. Chem. Phys.* **41**, 133 (1964).
86. A. L. Fetter and J. D. Walecka, *Quantum Theory of Many-Particle Systems*, McGraw-Hill, New York, 1971, Chapter 9.
87. D. Thirumalai and B. J. Berne, *J. Chem. Phys.* **79**, 5029 (1983).
88. P. Nielaba, J. L. Lebowitz, H. Spohn, and J. L. Valles, *J. Stat. Phys.* **55**, 745 (1989).
89. Y.-C. Chen, J. L. Lebowitz, and P. Nielaba, *J. Chem. Phys.* **91**, 340 (1989).
90. A. L. Nichols, III and D. Chandler, *J. Chem. Phys.* **87**, 6671 (1987).
91. E. C. Behrman, G. A. Jongeward, and P. G. Wolynes, *J. Chem. Phys.* **79**, 6277 (1983).
92. E. C. Behrman, G. A. Jongeward, and P. G. Wolynes, *J. Chem. Phys.* **83**, 668 (1985).
93. E. C. Behrman and P. G. Wolynes, *J. Chem. Phys.* **83**, 5863 (1985).
94. B. J. Berne and G. D. Harp, *Adv. Chem. Phys.* **17**, 63 (1970).
95. D. Thirumalai and B. J. Berne, *J. Chem. Phys.* **81**, 2512 (1984).
96. W. H. Miller, S. D. Schwartz, and J. W. Tromp, *J. Chem. Phys.* **79**, 4889 (1983).
97. K. Yamashita and W. H. Miller, *J. Chem. Phys.* **82**, 5475 (1985).
98. D. Thirumalai, B. C. Garrett, and B. J. Berne, *J. Chem. Phys.* **83**, 2972 (1985).
99. J. D. Doll, T. L. Beck, and D. L. Freeman, *J. Chem. Phys.* **89**, 5753 (1988).
100. G. Baym and N. D. Mermin, *J. Math. Phys.* **2**, 232 (1961).
101. C. J. Thompson, *Classical Equilibrium Statistical Mechanics*, Clarendon, Oxford, 1988, pp. 91–95.
102. R. E. Rosensweig, *Sci. Am.* **247**, 136, October 1982.
103. T. DeSimone and R. M. Stratt, *Phys. Rev. B* **32**, 1537 (1985).
104. R. M. Stratt, *Phys. Rev. B* **33**, 1921 (1986).
105. H. C. Andersen, D. Chandler, and J. D. Weeks, *Adv. Chem. Phys.* **34**, 105 (1976).
106. These problems may only be "MSA-like" because of the omission of h_0 bonds in the chain sum. Nonetheless, the analytical solutions to the MSA given in the literature can still be used if the symmetry of the problem is appropriate (as with dipolar fluids, Refs. 24 and 25, or certain mixture problems, *e.g.*, E. Waisman, *J. Chem. Phys.* **59**, 495 (1973)). The implications for the calculation of the excess internal energy are discussed in the appendix to Ref. 61.
107. D. Chandler, Y. Singh, and D. M. Richardson, *J. Chem. Phys.* **81**, 1975 (1984).

EQUILIBRIUM AND DYNAMICAL FOURIER PATH INTEGRAL METHODS

J. D. DOLL

Department of Chemistry, Brown University, Providence, Rhode Island

DAVID L. FREEMAN

Department of Chemistry, University of Rhode Island, Kingston, Rhode Island

and

THOMAS L. BECK

Department of Chemistry, University of Cincinnati, Cincinnati, Ohio

CONTENTS

I. INTRODUCTION

One of the principal tools available for the study of the properties of many-particle systems at nonzero temperatures is the Monte Carlo method.[1] Monte Carlo methods were introduced for statistical mechanics over 30 years ago,[2] and they have had a major impact on our understanding of condensed matter. Much of our current understanding of liquids, for example, has evolved directly from computer simulations utilizing Monte Carlo and molecular dynamics approaches. These numerical methods provide a tool whereby microscopic statistical mechanical theories can be implemented without the need for introducing uncontrolled approximations. Beyond their ability to produce particular results of interest for specific systems, these methods frequently can provide the level of insight into complex, many-body phenomenology that is required for the construction of physically sound, analytic theories.

As computational capabilities evolved over the past decades, it became clear that the original Monte Carlo algorithms that were developed strictly for classical systems could be extended to the study of quantum phenomena. Numerical path integral methods[3] and Green's function Monte Carlo approaches[4] have been developed that permit the examination of finite temperature and ground-state equilibrium properties of condensed matter systems. More recently, "real time" Monte Carlo methods[5-9] have been introduced in an attempt to extend statistical approaches to quantum dynamical problems. The ability to provide a common language for the description and treatment of such a broad range of physical phenomena and systems is a significant feature of the present methods.

The focus of the current article is path integral approaches to computational statistical mechanics. An important goal of the presentation is to make these methods available to those not previously familiar with the techniques. To make the presentation of manageable proportions, we limit the language of the present article to the Fourier path integral method.[10-13] It is the Fourier method with which we are personally most familiar. Reviews of an alternate approach, the discretized path integral method, have appeared elsewhere.[14] In addition to describing the formal structure of the theory, we have also illustrated developments with a common application, the one-dimensional quartic oscillator, for pedagogical purposes. The quartic oscillator is sufficiently simple for careful numerical study and can serve as a proto-type for general, anharmonic systems. Equilibrium methods are developed in Section II of this chapter, and the extensions to dynamical problems are contained in Section III. For simplicity, all formal developments are limited to one-dimensional systems, with the multidimensional extensions being indicated in the Appendix. In addition to the quartic oscillator examples mentioned above, sections include examples of many-body applications taken from liquid state and cluster chemistry and physics.

II. EQUILIBRIUM METHODS

In this section we develop the formal and numerical methods necessary to apply path integral techniques to the evaluation of equilibrium properties of macroscopic interacting many-body systems using the Fourier approach. It is first worth noting that the development will be limited to calculations utilizing the canonical ensemble. This is in contrast to classical Monte Carlo methods where a variety of ensembles can be and have been used. Path integral methods have employed the canonical ensemble by virtue of the formal correspondence between the quantum density matrix and the quantum propagator. Although some efforts have been made to expand the methods beyond the canonical ensemble,[15] these attempts will not be considered here. In what follows we also limit the discussion to quantum systems where the effects of particle exchange can be ignored. For bosonic systems, particle exchange contributions have been included in studies of liquid helium,[16] and attempts to find suitable algorithms for including exchange effects for fermions[17] are extensive. Since our concerns are in bulk systems of chemical interest, the effects of particle exchange will nearly always be negligible, and we ignore the contributions here. Of course, the standard factors of $(1/N!)$ characteristic of Boltzmann statistics do need to be included in any statistical mechanical calculation. For the sake of simplicity, we limit the discussion that follows to one-dimensional systems. The extension to many-particle systems in three dimensions is straightforward. For completeness, we outline the extensions for the many-particle, three-dimensional case in the Appendix. For the sake of clarity we have divided the present section into three subsections. In the first subsection we present the formal development. In the second, we apply the techniques developed in the first to a quartic potential. This same potential function will be used throughout this review to illustrate the methods we develop. In the final subsection we shall review applications of the methods to some physical systems of interest in cluster and liquid state chemistry and physics.

A. Formal Development

1. Introduction

The method to be developed in this review is the Fourier path integral Monte Carlo approach[10-13] to statistical mechanics. In fact there have been two principal approaches to quantum statistical mechanics using path integrals within a Monte Carlo context. The other approach has often been termed the discretized path integral method, and it has been the subject of a recent review.[14] The discretized approach provides a powerful technique for numerical applications as well as a convenient starting point for the development of path integral methods. We make no effort to provide a complete review of discretized methods. We merely use the formalism as a vehicle for developing the Fourier approach. Before beginning the actual development, it is useful to review Monte Carlo approaches to classical statistical mechanics and textbook approaches to quantum statistical mechanics. By reviewing this material, the utility of path integral methods can be realized.

In classical statistical mechanics for a system of defined temperature T and volume V, it is assumed that the measured value of a mechanical property P (e.g., the energy) is given by the expectation of the property over the Boltzmann distribution defined by the system Hamiltonian H. We use $\langle P \rangle$ to denote such an average so that

$$\langle P \rangle = \frac{\int d\Gamma \exp(-\beta H) P(\Gamma)}{\int d\Gamma \exp(-\beta H)}, \qquad (2.1)$$

where Γ represents the collective set of coordinates x and p, the position and momentum of the particle, respectively, and $\beta = 1/k_B T$ where k_B is the Boltzmann constant. Monte Carlo methods, particularly in the form introduced by Metropolis et al.,[2] are suited to the numerical evaluation of Eq. (2.1) especially in the many-particle case. The Metropolis Monte Carlo methods are designed to evaluate numerically integrals of the generic form

$$I = \frac{\int d\mathbf{r} \, \rho(\mathbf{r}) f(\mathbf{r})}{\int d\mathbf{r} \, \rho(\mathbf{r})}, \qquad (2.2)$$

where **r** represents any multidimensional collective coordinate and ρ is any positive definite weight function. In Metropolis Monte Carlo methods, Eq. (2.2) is evaluated by performing a random walk to generate a set of points distributed as $\rho(\mathbf{r})$. These points can then be used to estimate Eq. (2.2) as

$$I = \frac{1}{N} \sum_{i=1}^{N} f(\mathbf{r}_i), \tag{2.3}$$

where $\mathbf{r}_i$ is the ith point generated by the random walk. The estimate can be shown to become exact in the limit that the number of points N is taken to infinity. The details of Metropolis Monte Carlo methods have been reviewed many times,[18] and we make no effort to discuss them here. However, we wish to stress that integrals of the form of Eq. (2.2) can be evaluated in a well-prescribed fashion. As discussed elsewhere,[1] the chief advantage of Monte Carlo approaches is the insensitivity of the method to the dimensionality of the integration to be performed. This feature makes Monte Carlo methods very popular in statistical mechanics, where the dimensionality of integrations of the form of Eq. (2.1) can be very large. We shall restate this point later when we indicate that quantum statistical mechanical averages can be evaluated from integrals of the form of Eq. (2.2) by introducing additional degrees of freedom into the problem.

In quantum statistical mechanics for systems at a defined temperature T and volume V it is assumed that for every property there exists a Hermitian operator P such that

$$\langle P \rangle = \frac{\text{tr}(\rho P)}{\text{tr}(\rho)}, \tag{2.4}$$

where ρ is the quantum density operator defined by

$$\rho = \exp(-\beta H), \tag{2.5}$$

and H is the system Hamiltonian operator. Standard textbook expressions can be derived from Eq. (2.4) if the traces are evaluated in energy representation. For example, in energy representation the thermodynamic internal energy $\langle E \rangle$ is given by

$$\langle E \rangle = \frac{\sum_n E_n e^{-\beta E_n}}{\sum_n e^{-\beta E_n}}. \tag{2.6}$$

Although Eq. (2.6) is of considerable import in analytic evaluations of the energy for textbook problems amenable to such solutions, the expression is nearly useless for the evaluation of the energy of a complex interacting many-body system. Direct application of Eq. (2.6) would require the knowledge of the entire energy spectrum of the system. Such information is available for only a few idealized systems. This limitation has made simulation studies of quantum many-body systems impossible until recently.

An alternative to the evaluation of properties in energy representation, resulting in expressions like Eq. (2.6), utilizes the invariance of quantum mechanical traces to unitary transformations. If we evaluate Eq. (2.4) in coordinate representation, we obtain the expression

$$\langle P \rangle = \frac{\int dx \langle x| P \exp(-\beta H)|x \rangle}{\int dx \langle x| \exp(-\beta H)|x \rangle}. \tag{2.7}$$

Using closure with respect to a complete set of coordinate states, Eq. (2.7) becomes

$$\langle P \rangle = \frac{\int dx\, dx' \langle x| P |x' \rangle \langle x'| \exp(-\beta H)|x \rangle}{\int dx \langle x| \exp(-\beta H)|x \rangle}. \tag{2.8}$$

In the case that P is an operator diagonal in the coordinates, Eq. (2.8) takes the form

$$\langle P \rangle = \frac{\int dx\, P(x) \langle x| \exp(-\beta H)|x \rangle}{\int dx \langle x| \exp(-\beta H)|x \rangle}. \tag{2.9}$$

Equation (2.9) is in the form of a Metropolis Monte Carlo average if we identify the quantum density with the diagonal matrix elements of the density operator. We now introduce the notation for the density matrix, $\rho(x, x'; \beta)$

$$\rho(x, x'; \beta) = \langle x'| \exp(-\beta H)|x \rangle. \tag{2.10}$$

Clearly, we can transform the problem of quantum statistical mechanics into one of a classical form if we can discover a method of evaluating the quantum density matrix elements. Such a construction is accomplished by employing Feynman path integrals.[19] Before demonstrating this construction, it is of interest to note a formal analogy between the density matrix elements and the quantum mechanical propagator. Recalling that the quantum propagator is defined by

$$K(x, x'; t) = \langle x'| \exp(-iHt/\hbar)|x \rangle, \tag{2.11}$$

we note the quantum propagator is formally identical to the density matrix evaluated at an imaginary time $t = -i\beta\hbar$. This formal analogy will be exploited in the development of path integral approaches to statistical mechanics. As we shall demonstrate later, at least in the case of the energy, some operators not diagonal in the coordinates can also be transformed into the form of a Metropolis Monte Carlo average.

2. Path Integral Methods

We now proceed to the development of path integral expressions for the quantum density matrix. We begin by rewriting Eq. (2.10) in the form

$$\rho(x, x'; \beta) = \langle x' | [\exp(-\beta H/M)]^M | x \rangle, \tag{2.12}$$

where M is an integer. Between each of the M factors we can introduce a complete set of coordinate states so that Eq. (2.12) becomes

$$\rho(x, x'; \beta) = \int dx_1 \, dx_2 \ldots dx_{M-1} \langle x' | e^{-\beta H/M} | x_{M-1} \rangle$$
$$\times \langle x_{M-1} | e^{-\beta H/M} | x_{M-2} \rangle \ldots \langle x_1 | e^{-\beta H/M} | x \rangle. \tag{2.13}$$

Each density matrix element appearing on the right-hand side of Eq. (2.13) is itself in the form of a propagator evaluated at imaginary time $t = -i\beta\hbar/M$. Such density matrix elements are evaluated at the equivalent of a high temperature β/M, and a variety of analytic approximations are available for the density matrix elements[20] which become exact in the limit that $M \to \infty$. Any of these approximate expressions can be used in the development, and we choose one of the simplest for conveinence. To deduce a high temperature form for the density matrix elements, we begin by noting that the density matrix elements obey the Bloch equation

$$-\partial\rho/\partial\beta = H\rho. \tag{2.14}$$

Equation (2.14) follows from the fact that the propagator satisfies the Schroedinger equation, and the Bloch equation can be seen to be the Schroedinger equation in imaginary time. As a boundary condition on the Bloch equation, we impose

$$\rho(x, x'; \beta = 0) = \delta(x - x'). \tag{2.15}$$

The Bloch equation can be solved analytically for a free particle and for the harmonic oscillator. Here we need the free-particle solution. Using the free particle Hamiltonian

$$H = -\frac{\hbar^2}{2m} \frac{d^2}{dx^2}, \tag{2.16}$$

it is elementary to show by substitution that the free-particle solution to the Bloch equation is given by

$$\rho_{fp}(x, x'; \beta) = \left[\frac{m}{2\pi\hbar^2\beta} \right]^{1/2} \exp\left[-\left(\frac{m}{2\hbar^2\beta} \right)(x - x')^2 \right]. \qquad (2.17)$$

For high temperatures (β small), perturbation theory can be used[21] to show that if a potential $V(x)$ is added to the Hamiltonian in Eq. (2.16), the density matrix is given approximately by

$$\lim_{\beta \to 0} \rho(x, x'; \beta) = \left[\frac{m}{2\pi\beta\hbar^2} \right]^{1/2} \exp\left[-\frac{m(x - x')^2}{2\beta\hbar^2} - \beta V(x) \right]. \qquad (2.18)$$

As noted in Eq. (2.18), the approximation becomes exact at infinite temperature. Consequently, for each matrix element appearing in Eq. (2.13), we can introduce the approximation

$$\rho\left(x, x'; \frac{\beta}{M} \right) \cong \left[\frac{Mm}{2\pi\beta\hbar^2} \right]^{1/2} \exp\left[-\frac{Mm(x - x')^2}{2\beta\hbar^2} - \frac{\beta}{2M} V(x) - \frac{\beta}{2M} V(x') \right]. \qquad (2.19)$$

Writing

$$\varepsilon = \beta\hbar/M, \qquad (2.20)$$

Eq. (2.13) becomes

$$\rho(x, x'; \beta) \cong \left[\frac{m}{2\pi\hbar\varepsilon} \right]^{(M-1)/2} \int dx_1 \int dx_2 \dots \int dx_{M-1} \exp\left\{ -\frac{m\varepsilon}{2\hbar} \left[\left(\frac{x' - x_{M-1}}{\varepsilon} \right)^2 \right.\right.$$
$$\left. + \left(\frac{x_{M-1} - x_{M-2}}{\varepsilon} \right)^2 + \dots + \left(\frac{x_1 - x}{\varepsilon} \right)^2 \right] \right) - \frac{\beta}{2M} V(x)$$
$$\left. - \frac{\beta}{M} V(x_1) - \frac{\beta}{M} V(x_2) - \dots - \frac{\beta}{M} V(x_{M-1}) - \frac{\beta}{2M} V(x') \right\}. \qquad (2.21)$$

Equation (2.21) is often called the discretized representation of path integrals and has been the starting point for a number of numerical algorithms developed for computational quantum statistical mechanics. As we stated before, a discussion of discretized methods has been given by Berne and Thirumalai,[14] and the interested reader can refer to that presentation. We choose instead to use Eq. (2.21) to develop the full path integral representation for the quantum density matrix. To proceed, we introduce the obvious notation in the limit that $M \to \infty$ or equivalently $\varepsilon \to 0$

$$\frac{x_n - x_{n-1}}{\varepsilon} = \dot{x}(u)\Big|_{n\varepsilon}, \tag{2.22}$$

so that we can write

$$\rho(x, x'; \beta) = \int Dx(u) \exp\{-S[x(u)]\}, \tag{2.23}$$

where

$$S[x(u)] = \frac{1}{\hbar} \int_0^{\beta\hbar} du \left[\frac{m\dot{x}(u)^2}{2} + V(x(u)) \right]. \tag{2.24}$$

In the path integral, Eq. (2.23), $\int Dx(u)$ implies a summation over all paths which connect x to x' in imaginary time $\beta\hbar$. The u integration in Eq. (2.24) is of the form of a classical action integral in imaginary time. We shall refer to such integrals of the generic form of Eq. (2.24) as "action integrals" where it is convenient. We can represent the notion of a path integral by showing a pictorial representation of Eq. (2.13) as given in Fig. 1. Each matrix element

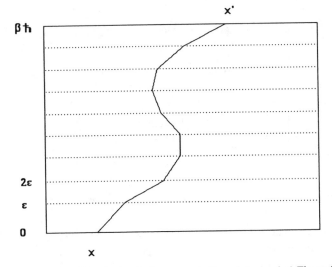

Figure 1. Shown is a discretized path that connects the points x and x'. The path begins at x at $u = 0$ and ends at x' at $u = \beta\hbar$ and is parameterized by its values at the intermediate time points $u_n = n\varepsilon$, where $\varepsilon = \beta\hbar/M$.

in Eq. (2.13) is a propagator in the complex time variable u which ranges from 0 to ε in the first matrix element, ε to 2ε in the second, and so on. In Fig. 1 each intermediate integration variable is represented as a point along the x axis, with the time variable u displayed along the y axis. According to the path integral prescription, the exact density matrix is found in the limit that $M \to \infty$ by summing all possible paths. As depicted in Fig. 1, these paths are generated by connecting all possible coordinates x_1 to x_{M-1} which are the integration variables in Eq. (2.21). Both differentiable and nondifferentiable paths are included.

3. Fourier Path Integral Methods

In the Fourier path integral method, the path integration in Eq. (2.23) is replaced by an ordinary Riemann integral. This is accomplished by writing each contributing path in a Fourier series and performing the path integration by ordinary integration over the Fourier coefficients. This approach was first introduced by Feynman and Hibbs[19] to obtain the analytic expression for the harmonic oscillator propagator. More recently the Fourier method has found applications in chemical reaction dynamics by Miller[22] as well as in the context of computational quantum statistical mechanics[10-13] which is the subject of this review.

To begin, as indicated we represent each path parametrically as a function of the time variable u using a Fourier series,

$$x(u) = x + (x' - x)u/\beta\hbar + \sum_{k=1}^{\infty} a_k \sin(k\pi u/\beta\hbar). \tag{2.25}$$

We then introduce Eq. (2.25) into Eq. (2.23). The kinetic energy u integrations can be performed analytically so that we obtain

$$\frac{\rho(x, x'; \beta)}{\rho_{fp}(x, x'; \beta)} = \frac{\int d\mathbf{a} \exp(-\sum_{k=1}^{\infty} a_k^2/2\sigma_k^2 - \beta\langle V\rangle_\mathbf{a})}{\int d\mathbf{a} \exp(-\sum_{k=1}^{\infty} a_k^2/2\sigma_k^2)}, \tag{2.26}$$

where

$$\sigma_k^2 = 2\beta\hbar^2/m(k\pi)^2, \tag{2.27}$$

and where the average of the potential energy along the path specified by Eq. (2.25) is given by

$$\langle V\rangle_\mathbf{a} = \frac{1}{\beta\hbar} \int_0^{\beta\hbar} du\, V(x(u)). \tag{2.28}$$

In later work we will sometimes find it convenient to express the results given above by transforming the u integration variable to the dimensionless time variable $u/\beta\hbar$, so that the range of integration is zero to one. For the present, however, we will retain the notation introduced above as a reminder of the physical origin of this temporal average. Equation (2.26) is in the form of an average of Boltzmann-like factors over the Fourier coefficients. This equation in conjunction with Monte Carlo methods can be used to compute the density matrix elements. Although the particle density by itself can be of interest, we shall gain further insight by using Eq. (2.26) to evaluate the expectation of an operator which depends only on the coordinates, so that only the diagonal elements of the density matrix are required. To be specific and for later use, we use Eq. (2.26) to evaluate the expectation of the potential energy in the canonical ensemble. It is first useful to introduce the notation

$$w(x, x', \mathbf{a}; \beta) = \rho_{fp}(x, x'; \beta) \exp\left(- \sum_{k=1}^{\infty} a_k^2/2\sigma_k^2 - \beta\langle V\rangle_\mathbf{a} \right), \qquad (2.29)$$

so that the expectation value of the potential energy can be written

$$\langle V\rangle = \frac{\int dx\, d\mathbf{a}\, w(x, x, \mathbf{a}; \beta) V(x)}{\int dx\, d\mathbf{a}\, w(x, x, \mathbf{a}; \beta)}. \qquad (2.30)$$

The average given above is over Cartesian and Fourier path variables. This average is distinguished from the temporal average over the time variable u for a specified quantum mechanical path by the subscript $\mathbf{a}$ in this latter result (compare Eqs. 2.28 and 2.30). The form of Eq. (2.30) is identical to a classical Monte Carlo average except that the dimensionality of the integrations is increased by the introduction of the Fourier coefficients as variables of integration. This introduction of auxiliary degrees of freedom in the quantum expressions is characteristic of path integral treatments in general. In the discretized formulation of path integral Monte Carlo, the form of Eq. (2.30) is the same except that Eq. (2.21) is used in place of $w(x, x, \mathbf{a}; \beta)$. In the discretized formulation, the auxiliary degrees of freedom are the integration variables x_1 to x_{M-1} rather than the Fourier coefficients. We do not discuss the question of the relative merits of the discretized and Fourier formulations for numerical study. Each has its regime of utility and each provides insight into the quantum many-body problem. Discussions of the discretized approach can be found in the review by Berne and Thirumalai[14] and the original papers by Barker[23] and by Chandler, Wolynes and co-workers.[24,25] In addition, the reader is referred to an interesting article by Coalson[26] on the formal connections between the approaches.

We now make two remarks concerning Eqs. (2.29) and (2.30). The first concerns the gaussian-like averages over the Fourier coefficients. These gaussian averages introduce σ_k as a natural length scale. This length scale is seen to grow as the temperature is decreased, and, like the thermal deBroglie wavelength, provides a measure of the extent of quantum contributions in the system. In fact, from Eq. (2.26) it is clear that

$$\lim_{\beta \to 0} \rho(x, x'; \beta) = \left[\frac{m}{2\pi\beta\hbar^2} \right]^{1/2} \exp\{ -\beta V(x) \}, \qquad (2.31)$$

which is just the classical density. The utility of using the gaussian widths as a length scale will be made clearer when we discuss the method of partial averaging.[27,28] The second remark concerns applications of the Fourier expressions in actual calculations. It is clear that the dimensionality of the integrations in Eq. (2.30) are infinite and must be truncated in actual numerical work. It is usual to truncate the number of Fourier coefficients included at some upper limit which we denote as k_{max}. With such truncations in mind we introduce the notation

$$\rho_{k_{max}}(x, x'; \beta) = \rho_{fp}(x, x'; \beta) \frac{\int d\mathbf{a} \exp(-\sum_{k=1}^{k_{max}} a_k^2/2\sigma_k^2 - \beta\langle V\rangle_\mathbf{a})}{\int d\mathbf{a} \exp(-\sum_{k=1}^{k_{max}} a_k^2/2\sigma_k^2)}, \qquad (2.32)$$

where in the truncated form we write

$$d\mathbf{a} = da_1 \, da_2 \dots da_{k_{max}}, \qquad (2.33)$$

and where the path specified by the truncated set of Fourier variables is given by

$$x_\mathbf{a}(u) = x + (x' - x)\frac{u}{\beta\hbar} + \sum_{k=1}^{k_{max}} a_k \sin(k\pi u/\beta\hbar). \qquad (2.34)$$

The time average of the potential in Eq. (2.32) is of the form of Eq. (2.28), except here the path is specified by Eq. (2.34). In the truncated form, we also introduce the notation

$$w_{k_{max}}(x, x', \mathbf{a}; \beta) = \rho_{fp}(x, x'; \beta)\exp\left(-\sum_{k=1}^{k_{max}} a_k^2/2\sigma_k^2 - \beta\langle V\rangle_\mathbf{a} \right) \qquad (2.35)$$

and write canonical averages of operators diagonal in the coordinate representation [e.g., $V(x)$] with the notation

$$\langle V \rangle_{k_{\max}} = \frac{\int dx\, d\mathbf{a}\, w_{k_{\max}}(x, x, \mathbf{a}; \beta) V(x)}{\int dx\, d\mathbf{a}\, w_{k_{\max}}(x, x, \mathbf{a}; \beta)}. \tag{2.36}$$

The exact quantum expectation is obtained in the limit that all Fourier coefficients are included, that is,

$$\lim_{k_{\max} \to \infty} \langle V \rangle_{k_{\max}} = \langle V \rangle \tag{2.37}$$

The utility of the Fourier method depends upon the rapidity with which the numerical results approach a limit as $k_{\max}$ is increased. We shall discuss that issue further in a later section where we examine some numerical results. For completeness, we now consider the evaluation of the kinetic energy.

4. Kinetic Energy Estimation

The kinetic energy requires more thought than the potential energy because the kinetic energy operator is not diagonal in coordinate representation. To make calculations of the kinetic energy practical in a Monte Carlo context, several approaches to the evaluation of its expectation have been developed. The first method we discuss evaluates the energy $\langle E \rangle$ in its entirety, rather than the kinetic energy alone. It utilizes the statistical mechanical expression

$$\langle E \rangle = -(\partial \ln Q / \partial \beta)_v \tag{2.38}$$

and is termed the "T method"[23,13] (T for temperature differentiation). The function Q appearing in Eq. (2.38) is the standard canonical partition function and is given by

$$Q(T, V, N) = \int dx\, \rho(x, x; \beta). \tag{2.39}$$

A second approach uses Eq. (2.8) with the kinetic energy operator introduced for P. The resulting expression for the average kinetic energy $\langle K \rangle$ is given by

$$\langle K \rangle = \frac{\int dx\, K\rho(x, x'; \beta)|_{x=x'}}{\int dx\, \rho(x, x; \beta)}, \tag{2.40}$$

where K is the kinetic energy operator. This approach has been termed the "H method"[12,13] (H for Hamiltonian operating on the density matrix). This expression is in the form of a Metropolis Monte Carlo average because x' is set to x after the differentiation. If we insert the truncated density matrix $\rho_{k_{\max}}(x, x'; \beta)$ given by Eq. (2.32) into Eq. (2.40), we obtain

$$\langle K \rangle_{k_{max}} = (1/2\beta) - \frac{\hbar^2}{2m}$$

$$\times \frac{\int dx\, da\, w_{k_{max}}(x, x, \mathbf{a}; \beta) \left\{ \left(\frac{1}{\hbar} \int_0^{\beta\hbar} du \left(1 - \frac{u}{\beta\hbar} \right) V'(x(u)) \right)^2 - \frac{1}{\hbar} \int_0^{\beta\hbar} du \left(1 - \frac{u}{\beta\hbar} \right)^2 V''(x(u)) \right\}}{\int dx\, da\, w_{k_{max}}(x, x, \mathbf{a}; \beta)}.$$

(2.41)

As we shall see, the T method has the virtue of simplicity. However, as we shall also see, the variance of the kinetic energy grows as the number of Fourier coefficients included is increased in the T method. This is potentially a serious limitation for highly quantum systems where many Fourier coefficients are required. In contrast, the variance of the kinetic energy is stable as the number of Fourier coefficients is increased in an H-method calculation. Consequently, the more complex expressions needed in the H method may be preferred in many cases. Because much of what we want to demonstrate concerns the dependence of the variance on the number of Fourier coefficients included, we work with the expressions for the density matrix truncated in the number of Fourier coefficients.

We begin by examining the energy in the T method. We introduce Eq. (2.32) and use Eq. (2.38) to obtain

$$\langle E \rangle_{k_{max}} = \frac{k_{max} + 1}{2\beta} + \frac{\int dx\, da\, w_{k_{max}}(x, x, \mathbf{a}; \beta) \left\{ \langle V \rangle_{\mathbf{a}} - \sum_{k=1}^{k_{max}} a_k^2 / 2\beta\sigma_k^2 \right\}}{\int dx\, da\, w_{k_{max}}(x, x, \mathbf{a}; \beta)}.$$

(2.42)

We can identify the first term on the right-hand side of Eq. (2.42) as well as the last term in the numerator with the kinetic energy in the T method. The remaining term approaches the exact thermodynamic average of the potential energy in the limit that $k_{max} \to \infty$. As we mentioned, the variance of the kinetic energy evaluated from Eq. (2.42) will grow as the number of Fourier coefficients included is increased. This can be seen by an evaluation of the energy of a free particle $[V(x) = 0]$ using Eq. (2.42). For a free particle, Eq. (2.42) takes the form

$$\langle E \rangle_{k_{max}} = \frac{k_{max} + 1}{2\beta} - \frac{\int dx\, da\, w_{k_{max}}(x, x, \mathbf{a}; \beta) \left\{ \sum_{k=1}^{k_{max}} a_k^2 / 2\beta\sigma_k^2 \right\}}{\int dx\, da\, w_{k_{max}}(x, x, \mathbf{a}; \beta)}.$$

(2.43)

We now set

$$\alpha_{k_{max}} = \sum_{k=1}^{k_{max}} \frac{a_k^2}{2\beta\sigma_k^2}$$

(2.44)

and define

$$B(s) = \left\langle \exp\left\{ -s \sum_{k=1}^{k_{max}} \frac{a_k^2}{2\sigma_k^2} \right\} \right\rangle_{k_{max}}. \tag{2.45}$$

Then

$$\langle \alpha_{k_{max}} \rangle_{k_{max}} = -\frac{1}{\beta}\left(\frac{\partial \ln B}{\partial s} \right)_{s=0} \tag{2.46}$$

and

$$\langle \delta\alpha_{k_{max}}^2 \rangle_{k_{max}} = \langle \alpha_{k_{max}}^2 \rangle_{k_{max}} - \langle \alpha_{k_{max}} \rangle_{k_{max}}^2 \tag{2.47}$$

$$= \frac{1}{\beta^2}\left(\frac{\partial^2 \ln B(s)}{\partial s^2} \right)_{s=0}. \tag{2.48}$$

Equation (2.45) can be evaluated analytically to give

$$\ln B(s) = -(k_{max}/2)\ln(1 + s), \tag{2.49}$$

so that

$$\langle \alpha_{k_{max}} \rangle_{k_{max}} = \frac{k_{max}}{2\beta} \tag{2.50}$$

and

$$\langle \delta\alpha_{k_{max}}^2 \rangle_{k_{max}} = \frac{k_{max}}{2\beta^2}. \tag{2.51}$$

From Eq. (2.51) it is clear that the standard deviation of the kinetic energy grows as the square root of the number of Fourier coefficients included in the T method. The expression for the kinetic energy estimator is very simple in the T method, and its evaluation is very easy to perform. However, the growth of the variance can be a serious problem for highly quantum systems or systems at low temperatures where many Fourier coefficients are required for accurate results. We add at this point that this growth in the variance is similar to difficulties in evaluating the kinetic energy in discretized formulations as discussed by Herman et al.[29]

We now examine the kinetic energy estimator in the H method. We shall find that the variance in this approach is only weakly dependent on the number of Fourier coefficients included, at least in the case of calculations on the linear one-dimensional harmonic oscillator. This is, however, at the expense of a considerably more complex kinetic energy estimator. We begin by noting that the density matrix elements (Eq. 2.32) can be evaluated analytically for the oscillator. The result is

$$\rho_{k_{\max}}(x, x'; \beta) = C \exp(-S(x, x', \beta)), \tag{2.52}$$

where C is a collection of system parameters but is independent of x and x'. $S(x, x', \beta)$ is given by

$$S(x, x', \beta) = \frac{m(x' - x)^2}{2\hbar^2\beta} + \frac{m\omega^2\beta}{6}(x^2 + x'^2 + xx')$$
$$- \frac{m}{\beta\hbar^2} \sum_{k=1}^{k_{\max}} \left(\frac{\beta\hbar\omega}{k\pi}\right)^2 \frac{(x - (-1)^k x')^2}{[1 + (k\pi/\beta\hbar\omega)^2]}. \tag{2.53}$$

From Eqs. (2.40) and (2.52) we thus have

$$\langle K \rangle_{k_{\max}} = \frac{\int dx(-\hbar^2/2m)(d^2\rho_{k_{\max}}(x, x'; \beta)/dx^2)_{x=x'}}{\int dx \, \rho_{k_{\max}}(x, x; \beta)}. \tag{2.54}$$

Carrying out the operations in Eq. (2.54) produces

$$\langle K \rangle_{k_{\max}} = \frac{\int dx \exp(-x^2/2\sigma^2) K_{k_{\max}}(x)}{\int dx \exp(-x^2/2\sigma^2)}, \tag{2.55}$$

where $K_{k_{\max}}(x)$ is defined by

$$K_{k_{\max}}(x) = \left(\frac{-\hbar^2}{2m}\right)[A^2 x^2 - B], \tag{2.56}$$

and where

$$\frac{1}{2\sigma^2} = \frac{m\omega^2\beta}{2} - \frac{m}{\beta\hbar^2} \sum_{k=1}^{k_{\max}} \left(\frac{\beta\hbar\omega}{k\pi}\right)^2 \frac{(1 - (-1)^k)^2}{[1 + (k\pi/\beta\hbar\omega)^2]}, \tag{2.57}$$

$$A = \frac{m\omega^2\beta}{2} - \frac{2m}{\beta\hbar^2} \sum_{k=1}^{k_{\max}} \left(\frac{\beta\hbar\omega}{k\pi}\right)^2 \frac{(1 - (-1)^k)}{[1 + (k\pi/\beta\hbar\omega)^2]}, \tag{2.58}$$

$$B = \frac{m}{\beta\hbar^2} + \frac{m\omega^2\beta}{3} - \frac{2m}{\beta\hbar^2} \sum_{k=1}^{k_{\max}} \left(\frac{\beta\hbar\omega}{k\pi}\right)^2 \frac{1}{[1 + (k\pi/\beta\hbar\omega)^2]}. \tag{2.59}$$

Tedious but straightforward algebra thus yields for the kinetic energy

$$\frac{\langle K \rangle_{k_{max}}}{\hbar\omega} = \frac{1}{2\beta\hbar\omega} + \frac{\beta\hbar\omega}{24} - \frac{1}{\beta\hbar\omega} \sum_{k=2,\text{even}}^{k_{max}} \left(\frac{\beta\hbar\omega}{k\pi}\right)^2 \frac{1}{[1 + (k\pi/\beta\hbar\omega)^2]}, \quad (2.60)$$

where the sum in Eq. (2.60) is over even values of the index k. If Eq. (2.55) were being evaluated by Monte Carlo methods, the associated statistical noise would be related to the root mean square (rms) fluctuations of the integrand $K_{k_{max}}(x)$. Since the probability distribution in Eq. (2.55) is a gaussian, this rms fluctuation can be easily evaluated. Explicitly,

$$\frac{\int dx \exp(-x^2/2\sigma^2)(K_{k_{max}}(x) - \langle K \rangle_{k_{max}})^2}{\int dx \exp(-x^2/2\sigma^2)} = \left(\frac{-\hbar^2}{2m}\right)^2 (2A^4\sigma^4). \quad (2.61)$$

Defining σ_{KE} as the square root the left-hand side of Eq. (2.61) and using Eqs. (2.57) and (2.58), we obtain

$$\frac{\sigma_{KE}}{\hbar\omega} = \frac{1}{2\sqrt{2}}\left\{\beta\hbar\omega/2 - \frac{1}{\beta\hbar\omega}\sum_{k=1}^{k_{max}}\left(\frac{\beta\hbar\omega}{k\pi}\right)^2(1 - (-1)^k)^2\Big/\left[1 + \left(\frac{k\pi}{\beta\hbar\omega}\right)^2\right]\right\}. \quad (2.62)$$

Terms in the sum in Eq. (2.62) decay as k^{-4} for large k, and thus σ_{KE}, unlike the corresponding T method result, does not grow in an unbounded way as the number of Fourier coefficients is increased. The explicit k_{max} dependence of σ_{KE} for the oscillator is shown in Fig. 2 for the highly quantum mechanical situation $\beta\hbar\omega = 100$. It is interesting to note from Fig. 2 that rather than increasing, as would be the case in the T method, σ_{KE} here actually *decreases*

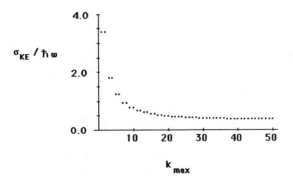

Figure 2. The rms width of the H-method kinetic energy estimator for the oscillator (Eq. 2.62) as a function of the number of Fourier coefficients (k_{max}).

as more quantum-mechanical variables are added to the problem. We conclude that the kinetic energy estimator in the H method, at least for the harmonic oscillator, has a well-defined variance as $k_{max} \to \infty$. Although we cannot present a formal proof, in all systems studied we have found no unbounded increase in the variance of the kinetic energy with respect to k_{max} when the H method is applied. We emphasize that although the growth in the variance of the kinetic energy in the T method with k_{max} is a serious limitation, the simplicity of its form and the ease of its evaluation may make the T method acceptable in some applications.

An alternate method for the evaluation of the kinetic energy utilizes the quantum mechanical virial theorem. The virial theorem states that

$$\langle K \rangle = \langle x \, dV/dx \rangle/2. \tag{2.63}$$

The advantage of this estimator is that the variance is independent of the number of Fourier coefficients included, and the evaluation of the expectation value is particularly simple. This form of kinetic energy estimator has been used in both discretized[23,29] and Fourier developments[12,13] of path integral methods. In applications using the Fourier approach, we have found the virial estimator to be particularly *unsuited* for calculations on systems described by strong, short-ranged repulsive potentials. Tests were performed on Lennard–Jones systems, and the convergence of the kinetic energy with respect to the number of Fourier coefficients included was poor, even at rather high temperatures. The principal difficulty seemed to be that the virial estimator is large in regions of configuration space rather far from those regions important to the Boltzmann weight. A numerical discussion of this point can be found elsewhere.[12] Because of the simplicity of the virial estimator, however, further exploration of this point seems warranted.

5. Partial Averaging

The success of the Fourier method depends upon the rapidity with which the calculated properties converge with respect to the number of Fourier coefficients included. The convergence rate is primarily a numerical issue, and a useful discussion is found in the next two subsections, where a number of example systems are presented. From the form of the gaussian factors in Eq. (2.32) and the associated length scale (i.e., σ_k as in Eq. 2.27), we can anticipate that as the temperature is decreased or as the system becomes increasingly quantum mechanical, an increasing number of Fourier coefficients will be required. This is indeed found to occur in practice. As discussed, each Fourier coefficient represents an additional auxiliary degree of freedom in the calculation, and the additional work arising from adding Fourier coefficients can be likened to adding additional particles in a classical calculation. As we add Fourier coefficients we can expect the numerical work to increase approximately linearly. In some cases the increase in work with the number of Fourier coefficients can be worse than linear because the u integrations, which are evaluated by numerical quadrature, can require increasing numbers of quadrature points with increasing k_{max}. With these remarks in mind, it is clear that finding a method for accelerating the convergence with respect to the number of included Fourier coefficients is very important. One such method has been termed *partial averaging*.[27,28] We shall discuss the partial average approach now.

The physical idea behind partial averaging is associated with the natural length scale, σ_k, discussed previously. As the Fourier index k increases, the length scale decreases. Beyond a sufficiently large Fourier index, k_{max}, we can imagine that the higher coefficients describe small fluctuations about the path fixed by the low-order Fourier coefficients, that is, those with index $k \leq k_{max}$. If these high-order fluctuations decouple sufficiently from the potential energy function, the mathematics of incorporating the effect of the higher Fourier coefficients becomes rather simple.

To develop partial averaging formally, it is useful to introduce a notation for performing integrations over the higher-order coefficients. For a general function of the Fourier coefficients g we write

$$\langle g \rangle_{\{a\}} = \frac{\int d\{a\} \exp[-\sum_{k_{max}+1}^{\infty} a_k^2/2\sigma_k^2] g(a)}{\int d\{a\} \exp[-\sum_{k_{max}+1}^{\infty} a_k^2/2\sigma_k^2]}, \tag{2.64}$$

where $\{a\}$ represents the set of Fourier coefficients $k > k_{max}$, that is,

$$d\{a\} = da_{k_{max}+1} da_{k_{max}+2} \ldots, \tag{2.65}$$

whereas

$$d\mathbf{a} = da_1 da_2 \ldots da_{k_{max}}. \tag{2.66}$$

Although the average in Eq. (2.64) removes explicit dependence on the $\{a\}$ coefficients, $\langle g \rangle_{\{a\}}$ retains a dependence on the low-order Fourier variables $\mathbf{a}$. Using the notation introduced in Eq. (2.64), Eq. (2.26) can be written

$$\frac{\rho(x, x'; \beta)}{\rho_{fp}(x, x'; \beta)} = \frac{\int d\mathbf{a} \exp[-\sum_{k=1}^{k_{max}} a_k^2/2\sigma_k^2] \langle \exp\{-1/\hbar \int_0^{\beta\hbar} du\, V(x(u))\} \rangle_{\{a\}}}{\int d\{a\} \exp[-\sum_{k=1}^{k_{max}} a_k^2/2\sigma_k^2]}. \tag{2.67}$$

It is now useful to introduce the Gibbs inequality which is often used in perturbative approaches to path integral methods,[19] namely,

$$\langle \exp(f) \rangle \geq \exp(\langle f \rangle), \tag{2.68}$$

where the average expressed in Eq. (2.68) can be taken over any distribution. Using the Gibbs inequality in Eq. (2.67), we can write

$$\rho(x, x'; \beta) \geq \rho_{k_{\max}}^{pa}(x, x'; \beta) \tag{2.69}$$

where

$$\frac{\rho_{k_{\max}}^{pa}(x, x'; \beta)}{\rho_{fp}(x, x'; \beta)} = \frac{\int d\mathbf{a} \exp\{-\sum_{k=1}^{k_{\max}} a_k^2/2\sigma_k^2 - 1/\hbar \int_0^{\beta\hbar} du \, V_{\mathrm{eff}}(x_\mathbf{a}(u), u)\}}{\int d\mathbf{a} \exp\{-\sum_{k=1}^{k_{\max}} a_k^2/2\sigma_k^2\}}, \tag{2.70}$$

and where $x_\mathbf{a}(u)$ is defined by Eq. (2.34).

In Eq. (2.70) we have defined the effective potential V_{eff} by

$$V_{\mathrm{eff}}(x_\mathbf{a}(u), u) = \left\langle V\left(x_\mathbf{a}(u) + \sum_{k_{\max}+1}^{\infty} a_k \sin(k\pi u/\beta\hbar) \right) \right\rangle_{\{\mathbf{a}\}}. \tag{2.71}$$

Working expressions for the effective potential can be developed from Eq. (2.71) using gaussian transform methods. Introducing explicit forms into Eq. (2.71) we obtain

$$V_{\mathrm{eff}}(x_\mathbf{a}(u), u) = \frac{\int d\{\mathbf{a}\} \exp[-\sum_{k_{\max}+1}^{\infty} a_k^2/2\sigma_k^2] V(x_\mathbf{a}(u) + \sum_{k_{\max}+1}^{\infty} a_k \sin(k\pi u/\beta\hbar))}{\int d\{\mathbf{a}\} \exp[-\sum_{k_{\max}+1}^{\infty} a_k^2/2\sigma_k^2]}, \tag{2.72}$$

Since V on the right-hand side of Eq. (2.72) depends only on a linear combination of gaussian variables, the infinite dimensional gaussian average can be reduced to an average over a single degree of freedom. To see this, we rewrite Eq. (2.72) as

$$V_{\mathrm{eff}}(x_\mathbf{a}(u), u)$$

$$= \frac{\int d\{\mathbf{a}\} \, dp \exp[-\sum_{k_{\max}+1}^{\infty} a_k^2/2\sigma_k^2] V[x_\mathbf{a}(u) + p] \delta[p - \sum_{k_{\max}+1}^{\infty} a_k \sin(k\pi u/\beta\hbar)]}{\int d\{\mathbf{a}\} \exp[-\sum_{k_{\max}+1}^{\infty} a_k^2/2\sigma_k^2]}. \tag{2.73}$$

Introducing a Fourier representation of the δ function gives

$$V_{\mathrm{eff}}(x_\mathbf{a}(u), u)$$

$$= \frac{\int d\{\mathbf{a}\} \, dp(dK/2\pi) \exp[-\sum_{k_{\max}+1}^{\infty} a_k^2/2\sigma_k^2] V[x_\mathbf{a}(u) + p] \exp[iK(p - \sum_{k_{\max}+1}^{\infty} a_k \sin(k\pi u/\beta\hbar))]}{\int d\{\mathbf{a}\} \exp[-\sum_{k_{\max}+1}^{\infty} a_k^2/2\sigma_k^2]}. \tag{2.74}$$

The integrations in Eq. (2.74) over the Fourier coefficients and over the Fourier index K are gaussians and can be performed analytically. After these integrations are performed, the result is

$$V_{\text{eff}}(x_a(u), u) = \frac{1}{\sqrt{2\pi}\sigma(u)} \int_{-\infty}^{\infty} dp \exp\left[-\frac{p^2}{2\sigma(u)^2}\right] V(x_a(u) + p), \quad (2.75)$$

with

$$\sigma^2(u) = \sum_{k_{\max}+1}^{\infty} \sigma_k^2 \sin^2\left(\frac{k\pi u}{\beta\hbar}\right) \quad (2.76)$$

$$= \frac{\beta\hbar^2}{m}\left[\frac{u}{\beta\hbar}\left(1 - \frac{u}{\beta\hbar}\right) - \sum_{k=1}^{k_{\max}} \sigma_k^2 \sin^2\left(\frac{k\pi u}{\beta\hbar}\right)\right] \quad (2.77)$$

From the mathematical development shown above, we can observe a number of points. The first is that the partial averaged density matrix is a lower bound to the true density matrix (Eq. 2.69). The second is that the effective potential needed in partial average calculations is the gaussian transform of the system potential. When the gaussian transform is analytic, the construction of the effective potential is elementary, and the introduction of partial averaging is no more complex than the original primitive Fourier algorithm. When the gaussian transform is not analytic but exists, an approximate effective potential can be constructed by expanding the system potential appearing on the right-hand side of Eq. (2.75) in a Taylor series about the point $p = 0$. The resulting expansion of the effective potential then takes the form

$$V_{\text{eff}}(x_a(u), u) = V(x_a(u)) + \frac{1}{2}V''(x_a(u))\sigma^2(u) + \cdots. \quad (2.78)$$

This expression of the effective potential in a Taylor series can be useful even when the gaussian transform of the system potential does not exist in a formal sense. By replacing the effective potential in Eq. (2.70) by Eq. (2.78) for cases when the gaussian transform does not exist, the result is a well-defined approximation to the true density matrix, which becomes exact in the limit that $k_{\max}$ becomes infinite. This behavior is not different from the primitive Fourier algorithm, which also gives an approximate result when a finite number of Fourier coefficients are included. The objective of any of the forms of partial averaging is improved convergence characteristics of the density matrix and the calculated properties with respect to $k_{\max}$. We shall see that this objective is realized in actual calculations. There are classes of potentials for which partial averaging is not appropriate in any form (e.g., hard-sphere interactions).

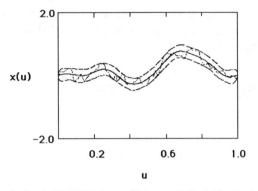

Figure 3. Plot of a "typical" 10-Fourier coefficient path (line). Superimposed on the plot is a 100-coefficient path (dashed line) that has the same low-order core. The long dashed line is the partial averaging envelope formed by adding $\pm 2[\sigma^2(u)]^{1/2}$ (cf. Eq. 2.77) to the 10-coefficient path.

Additional intuition concerning the method of partial averaging can be obtained by examining Fig. 3. In Fig. 3 the dashed line represents a typical path for the diagonal density matrix element when 100 Fourier coefficients are included. The set of Fourier coefficients, $\{a_k\}$, that describe this path were chosen at random from a gaussian distribution whose first and second moments are given by zero and $\{\sigma_k\}$, respectively, (cf. Eq. 2.27; $\beta = m = \hbar = 1$). The second path shown in Fig. 3 (solid line) is the low-order "core" of the first path obtained by setting all but the first 10 Fourier coefficients in the previous result equal to zero. The fluctuations of the higher-order path about the low-order core are thus a measure of the significance of the higher-order Fourier terms, terms assumed to be zero in the direct Fourier method. In partial averaging these terms are not assumed to be zero, but are approximated as free-particle fluctuations about a prescribed low-order path. Computationally, the effects of these fluctuations are included by replacing the value of the potential energy function at each point along the low-order path by a gaussian smear of the potential over a distribution of width $\sigma(u)$ (cf. Eq. 2.77). The bold dashed line in Fig. 3 depicts an envelope of $\pm 2\sigma(u)$ about the 10-coefficient path, an envelope that accurately describes the limits of the fluctuations of the higher-order path about its low-order core.

In deriving the expressions used in partial average calculations, our starting point was the Gibbs inequality. It is possible to gain further insight into partial averaging by identifying the partial average representation of the density matrix as the first term of a cumulant expansion.[30] For an arbitrary density function $P(x)$ and arbitrary function $f(x)$, the cumulant expansion takes the form

$$\int dx\, P(x)\exp\{f(x)\} = \exp\{\langle f \rangle + (1/2)(\langle f^2 \rangle - \langle f \rangle^2)$$

$$+ (1/3!)(\langle f^3 \rangle - 3\langle f^2 \rangle\langle f \rangle + 2\langle f \rangle^3) + \cdots\}, \qquad (2.79)$$

where the brackets $\langle\ \rangle$ represent an average over the density function $P(x)$. The chief advantage of expressing the partial averaged result as the first term of a cumulant expansion is the ability to improve the result by the inclusion of the higher-order terms given in Eq. (2.79). Within the cumulant approach, the possibility exists of finding better approximations than the truncated partial averaged algorithm. In practice, the calculation of the higher cumulants may prove to be more difficult than the savings realized by improvement in the accuracy of the approximate density matrix for finite k_{max}. Our numerical experience to date has been limited to the primitive partial averaged algorithm outlined above. However, further exploration of the higher-order cumulants does appear to be warranted. An additional advantage of the cumulant expansion concerns its utility in developing improved short-time propagators. We do not discuss such developments here, but refer the interested reader to the appropriate literature.[31,32]

B. The Quartic Oscillator

In this section we present results of Fourier path integral Monte Carlo studies for the calculation of the energy of a quartic oscillator, the potential energy function of which is given by

$$V = kx^4, \qquad (2.80)$$

where k is a proportionality coefficient having units of energy/length[4]. Our chief purpose in this subsection is to provide an illustration of the methods developed in the first subsection. As a secondary purpose it is hoped that the numerical results can provide a benchmark for those actually wishing to attempt Fourier path integral Monte Carlo calculations on a computer. We begin by focusing on Eqs. (2.28), (2.35), and (2.36) for the evaluation of the potential energy of the quartic oscillator in the approximation that only a finite number of Fourier coefficients are required for an accurate answer. As stated previously, Eq. (2.36) is in the form of a standard Metropolis Monte Carlo integral with both x and the Fourier coefficients as integration variables. The dimensionality of the Monte Carlo average is $k_{max} + 1$, and as in any Metropolis Monte Carlo calculation of a multidimensional integral, it can often prove necessary to make moves on each of the integration variables separately for each Monte Carlo pass. However, in integrations of the type given in Eq. (2.36) there are two alternatives for making moves in the Fourier coefficients. The first is the standard Metropolis procedure[2] where the Fourier coefficients are changed randomly with a uniform distribution. In that procedure, it may appear that the moves in the coordinates and the Fourier coefficients need to be made separately. In practice we have found that often the coordinate moves are more important. Consequently, we have moved the coordinates on each pass and simultaneously one associated Fourier coefficient. The chosen Fourier coefficient is picked at random. As an alternative, we note that the distribution of the Fourier coefficients in Eq. (2.35) is nearly gaussian, modified only by the dependence of the potential function in Eq. (2.35) on the Fourier coefficients. This suggests the alternate procedure where the coefficients are all moved with each attempted coordinate move, with the coefficients changed according to a gaussian distribution of width σ_k for each coefficient. We have experimented with both approaches, and the efficiencies of the two methods appears to be comparable. In the calculations on the quartic oscillator given below, we have chosen to move the Fourier coefficients according to the gaussian distribution.

To perform the calculation of the average of the potential energy, the second required quantity is the u integration appearing in the exponent on the right-hand side of Eq. (2.35). Although for the quartic oscillator this integral can be performed analytically, in general such u integrations must be evaluated numerically. The integrand is periodic with the full period being from $u = 0$ or $u = \beta\hbar$, so that equal-spaced trapezoidal rule quadrature is appropriate. In practice the number of quadrature points chosen is that required for accuracy in the final result to be within the statistical noise of the calculation. The determination of the optimal number of points is somewhat important, because the number of points determines the ultimate efficiency of the calculation. The actual number of quadrature points required depends upon circumstances and should be tested with a small number of Monte Carlo points for each new system attempted. A study of the dependence of the results on both the number of quadrature points and the number of Fourier coefficients included will be discussed below.

We shall also present results of the evaluation of the kinetic energy using both the H method and the T method. As is evident from Eq. (2.42), the evaluation of the kinetic energy using the T method is straightforward, and we shall not discuss it further here. The results will appear below. From Eq. (2.41) it is clear that in the case of the H method there are further numerical issues. In particular, the integrand contain two terms which must be evaluated with additional u integrations. We have found that the requirements for the u integrations are similar to the density, and in the results presented below the same number of quadrature points was used in all the u integrations, whether in the integrand or in the density.

The particular units chosen for this study were picked somewhat arbitrarily.

TABLE I
The Potential and Kinetic Energies of the Quartic Oscillator at $\beta\Delta E_{01} = 4$ for $k_{max} = 64$
as a Function of the Number of Quadrature Points Included in the u Integrations

N_{quad}	$\langle V \rangle / \Delta E_{01}$	$\langle K \rangle_H / \Delta E_{01}$	$\langle K \rangle_T / \Delta E_{01}$
2	0.0904 ± 0.0006	0.181 ± 0.009	0.181 ± 0.002
4	0.1139 ± 0.0008	0.234 ± 0.005	0.229 ± 0.002
8	0.129 ± 0.001	0.250 ± 0.003	0.251 ± 0.002
16	0.1322 ± 0.0007	0.266 ± 0.002	0.259 ± 0.002
32	0.138 ± 0.001	0.262 ± 0.002	0.262 ± 0.004
64	0.137 ± 0.001	0.272 ± 0.002	0.262 ± 0.002
128	0.137 ± 0.001	0.270 ± 0.003	0.261 ± 0.002
256	0.137 ± 0.001	0.270 ± 0.002	0.269 ± 0.002

The coefficient of the quartic potential k was set to be 1.00 au and the mass was chosen to be the mass of a hydrogen atom. The temperatures used are expressed in units of the energy of the ground to first excited state transition energy, ΔE_{01}. The three temperatures were chosen to be $\beta\Delta E_{01} = 1$, 2, and 4.

As indicated, the number of quadrature points chosen to evaluate the u integrations were taken to be that necessary to attain convergence in the final results to within the statistical noise of the calculation. In Table I we show the values of the kinetic and potential energies as a function of the number of quadrature points included [N_{quad}] when 64 Fourier coefficients were included in the calculations. The temperature used in Table I was $\beta\Delta E_{01} = 4$ and the energies are expressed in units of the 0–1 transition energy. As discussed above, equal-spaced trapezoidal rule quadrature was used for the u integrations. The error bars listed in the table are single standard deviation error bars resulting from an evaluation using one million Monte Carlo points. From the table, under these conditions convergence was reached when 32 quadrature points were included. In general the number of required quadrature points can be expected to grow as the temperature is lowered or when more Fourier coefficients are included. In the results discussed below, 64 quadrature points were found to be satisfactory for all conditions studied for the quartic oscillator at the calculated temperatures reported.

In Table II we give the average potential and kinetic energies as a function of the number of Fourier coefficients included for the three computed temperatures when the primitive Fourier algorithm is applied. The kinetic energy results were obtained both with the H method $\langle K \rangle_H$ and the T method, $\langle K \rangle_T$. In Table III the same quantities are given when partial averaging is included. For the quartic oscillator, the partial averaged potential is analytic and given by

TABLE II
The Potential and Kinetic Energy of the Quartic Oscillator as a Function of k_{max} in Units of ΔE_{01} Without the Inclusion of Partial Averaging[a]

$\beta\Delta E_{01}$	k_{max}	$\langle V \rangle/\Delta E_{01}$	$\langle K \rangle_H/\Delta E_{01}$	$\langle K \rangle_T/\Delta E_{01}$
1	1	0.310 ± 0.001	0.578 ± 0.001	0.516 ± 0.002
	2	0.301 ± 0.001	0.572 ± 0.001	0.538 ± 0.002
	4	0.296 ± 0.001	0.571 ± 0.001	0.552 ± 0.002
	8	0.290 ± 0.001	0.572 ± 0.001	0.566 ± 0.004
	16	0.288 ± 0.001	0.572 ± 0.001	0.568 ± 0.004
	∞	0.2861	0.5722	
2	1	0.208 ± 0.001	0.359 ± 0.001	0.2724 ± 0.0004
	2	0.193 ± 0.001	0.353 ± 0.001	0.299 ± 0.001
	4	0.188 ± 0.001	0.347 ± 0.001	0.318 ± 0.001
	8	0.183 ± 0.001	0.347 ± 0.002	0.331 ± 0.002
	16	0.178 ± 0.001	0.350 ± 0.001	0.341 ± 0.002
	32	0.1769 ± 0.0005	0.353 ± 0.002	0.350 ± 0.004
	∞	0.1753	0.3506	
4	1	0.166 ± 0.001	0.290 ± 0.002	0.1524 ± 0.0002
	2	0.148 ± 0.001	0.264 ± 0.002	0.1859 ± 0.0003
	4	0.151 ± 0.002	0.259 ± 0.003	0.2134 ± 0.0004
	8	0.143 ± 0.001	0.263 ± 0.003	0.2371 ± 0.0008
	16	0.140 ± 0.001	0.261 ± 0.003	0.252 ± 0.001
	32	0.139 ± 0.001	0.267 ± 0.002	0.262 ± 0.002
	64	0.137 ± 0.001	0.270 ± 0.002	0.269 ± 0.002
	∞	0.1351	0.2702	

[a] The subscripts H and T on the kinetic energies represent results of calculations using the H and T methods respectively (see text).

$$V_{eff}(x(u), u) = k\{x^4 + 6x^2\sigma^2(u) + 3\sigma^4(u)\} \qquad (2.81)$$

From the tables the improved convergence characteristics of partial averaging are evident. To clarify this point, the average potential energy at $\beta\Delta E_{01} = 4$ is plotted as a function of k_{max} in Fig. 4. Both the partial averaged and primitive Fourier results are shown. The improved convergence characteristics of the partial averaged results are evident.

Using the results presented in Tables II and III, it is of interest to compare the relative utilities of the H and T methods for kinetic energy evaluations. At high temperatures, the H method appears to be preferable, because the kinetic energy obtained from the H method is more rapidly converged than the T method, and because the variance is stable. The rapidity of the convergence at high temperatures is a result of the leading term of $1/2\beta$ in Eq. (2.41). The additional terms in Eq. (2.41) provide quantum corrections which are small at the higher temperatures. At low temperatures the convergence of the H and T methods for the kinetic energy is similar. It is of interest to note that

TABLE III
The Potential and Kinetic Energy of the Quartic Oscillator as a Function of k_{max} in Units of ΔE_{01} with Partial Averaging Included[a]

$\beta\Delta E_{01}$	k_{max}	$\langle V\rangle/\Delta E_{01}$	$\langle K\rangle_H/\Delta E_{01}$	$\langle K\rangle_T/\Delta E_{01}$
1	1	0.282 $\pm$ 0.001	0.577 $\pm$ 0.001	0.516 $\pm$ 0.001
	2	0.283 $\pm$ 0.001	0.572 $\pm$ 0.001	0.538 $\pm$ 0.002
	4	0.286 $\pm$ 0.001	0.572 $\pm$ 0.001	0.551 $\pm$ 0.002
	8	0.286 $\pm$ 0.001	0.571 $\pm$ 0.001	0.560 $\pm$ 0.002
	∞	0.2861	0.5722	
2	1	0.164 $\pm$ 0.001	0.366 $\pm$ 0.001	0.2718 $\pm$ 0.0005
	2	0.168 $\pm$ 0.001	0.352 $\pm$ 0.001	0.301 $\pm$ 0.001
	4	0.174 $\pm$ 0.001	0.351 $\pm$ 0.001	0.3203 $\pm$ 0.001
	8	0.176 $\pm$ 0.001	0.352 $\pm$ 0.001	0.331 $\pm$ 0.001
	16	0.175 $\pm$ 0.001	0.352 $\pm$ 0.001	0.349 $\pm$ 0.002
	∞	0.1753	0.3506	
4	1	0.110 $\pm$ 0.001	0.336 $\pm$ 0.003	0.1563 $\pm$ 0.0004
	2	0.114 $\pm$ 0.001	0.285 $\pm$ 0.002	0.1960 $\pm$ 0.0003
	4	0.129 $\pm$ 0.001	0.277 $\pm$ 0.003	0.2221 $\pm$ 0.0004
	8	0.135 $\pm$ 0.001	0.268 $\pm$ 0.003	0.2434 $\pm$ 0.0007
	16	0.135 $\pm$ 0.001	0.272 $\pm$ 0.003	0.2558 $\pm$ 0.0008
	32	0.134 $\pm$ 0.001	0.274 $\pm$ 0.002	0.266 $\pm$ 0.001
	64	0.134 $\pm$ 0.001	0.265 $\pm$ 0.003	0.266 $\pm$ 0.002
	∞	0.1351	0.2702	

[a] The subscripts H and T on the kinetic energies represent results of calculations using the H and T methods respectively (see text).

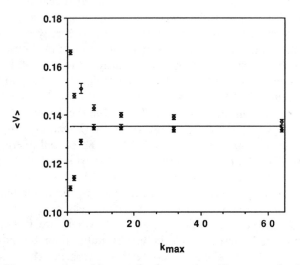

Figure 4. The average potential energy for the quartic oscillator example of Section II as a function of the number of Fourier coefficients in the calculation. The potential energy is in units of the ground to first vibrational state energy spacing, ΔE_{01}. The temperature for this example was $\beta\Delta E_{01} = 4.0$. The open (solid) squares are the results without (with) partial averaging and the exact value is indicated by the line.

TABLE IV

The Total Energy as a Function of the Number of Fourier Coefficients at a Temperature of $\beta\Delta E_{01} = 4$ using the H Method (E_H) and the Two T-Method Estimators (E_{T1} and E_{T2}) discussed in the text

k_{max}	$E_H/\Delta E_{01}$	$E_{T1}/\Delta E_{01}$	$E_{T2}/\Delta E_{01}$
1	0.456 ± 0.002	0.318 ± 0.001	0.2280 ± 0.0003
2	0.412 ± 0.002	0.3335 ± 0.0008	0.2785 ± 0.0004
4	0.410 ± 0.004	0.361 ± 0.001	0.3200 ± 0.0009
8	0.406 ± 0.003	0.379 ± 0.001	0.3555 ± 0.0005
16	0.401 ± 0.003	0.393 ± 0.002	0.379 ± 0.002
32	0.406 ± 0.002	0.398 ± 0.003	0.388 ± 0.002
64	0.407 ± 0.002	0.405 ± 0.003	0.400 ± 0.003
∞	0.4053		

the variance of the H method results is higher than the T method for small values of k_{max}. Although the variance of the T method results is seen to grow with the number of Fourier coefficients included, by the time convergence is reached, the H and T methods have similar variances.

Further insight into the relative characteristics of the H and T methods can be obtained by examining the convergence of the *total* energy with respect to the number of Fourier coefficients included. As we indicated in the discussion immediately below Eq. (2.42), within the T method there are two separate possible estimators of the potential energy. Because the expectation value of $\langle V \rangle_a$ is the same as the expectation value of V when an infinite number of Fourier coefficients are included, the potential energy can be evaluated either by averaging over V or $\langle V \rangle_a$. For a finite number of Fourier coefficients it is unclear which of the two potential energy estimates will be better converged. In Table IV we list the total energy calculated for the quartic oscillator at a temperature of $\beta\Delta E_{01} = 4$ as a function of the number of Fourier coefficients included within the H method (E_H), within the T method with V as the potential energy estimator (E_{T1}), and within the T method with $\langle V \rangle_a$ as the potential energy estimator (E_{T2}). In Table IV no partial averaging was included in the calculations. From the results it is evident that the H estimator is significantly better converged than either of the T estimators. In fact, to within the statistical precision of the calculations, the H method results for the total energy are converged by $k_{max} = 4$, whereas neither T method total energy results are converged until $k_{max} = 32$. Furthermore, it is apparent that the bare potential energy V is a better estimator than the path averaged estimator $\langle V \rangle_a$. The total energy has better convergence characteristics than either the individual potential or the kinetic energies in the H method, suggesting some level of cancellation of errors between the kinetic and potential energy contributions. Although not shown, we have performed the same study on the quartic oscillator with partial averaging included. Again we have found the total energy in the H method to converge significantly faster than the T method. When partial averaging was included, the total energy in the H method was converged at $k_{max} = 2$. Although the kinetic energy estimator in the T method is appealing because of its simplicity, we have found the H method to give superior results in all calculations we have tested to date.

While the results of this section are of interest as an illustration of the methods, the true power of Fourier path integral methods only becomes clear by application to interacting many-particle systems. A review of such applications is given in the next section.

C. Many-Particle Applications

In the previous section we illustrated the Fourier path integral Monte Carlo method with example calculations on the one-dimensional quartic oscillator. In this final subsection on equilibrium methods we illustrate the techniques developed with results on several interacting many-particle systems. The Fourier path integral method has been used in a number of applications to liquid state and cluster chemistry and physics. The examples picked for this section were chosen both to illustrate the utility of the technique and to identify one important outcome of the development of methods to study many-body systems in the quantum domain. Before the evolution of Monte Carlo methods into quantum mechanics, physical effects dependent on the quantum nature of matter could only be studied for a relatively restricted class of model problems or when perturbative approaches[33-35] were appropriate. Because of the path integral formulation, we can now study physical phenomena which were heretofore inaccessible. One of the first applications of the Fourier method was to the determination of the thermodynamic properties of clusters.[36] Clusters were chosen both for their interest and because their vibrational degrees of freedom were expected to have important quantum contributions at least at low temperatures. Because classical Monte Carlo methods have proved to be an important technique with which to study clusters, it seemed natural to apply path integral Monte Carlo to the same set of systems. The original calculations were on argon clusters[36] so that comparisons with previous classical work[37] could be discussed. More recently, quantum Monte Carlo methods have been used to study cluster melting in both argon and neon.[38] As a final illustration, we will present some recent results on the properties of liquid helium at temperatures above the lambda point so that particle exchange effects can be ignored. The interest in this system lies with the availability of some excellent discretized path integral calculations[39] for comparison. Furthermore, we use the path integral methods to examine approximations introduced by Weeks, et al.[40] for classical fluids to assess the extent to which such approximations are appropriate in the

TABLE V
The Lennard–Jones Parameters Used in the Calculations
Reported in this Section

	$\varepsilon(K)$	$\sigma(A)$
He	10.22	2.556
Ne	35.60	2.749
Ar	119.4	3.405

quantum domain. As in the first two examples, such an assessment would be very difficult if not impossible without path integral methods.

In all of the three systems to be discussed in this subsection, it is assumed that the forces between the constituent atoms can be represented by a Lennard–Jones pair interaction; i.e., we concentrate on systems whose potential energy function is given by

$$V(\mathbf{r}_1, \mathbf{r}_2, \ldots, \mathbf{r}_N) = \sum_{i<j}^{N} v(|\mathbf{r}_i - \mathbf{r}_j|), \tag{2.82}$$

where

$$v(r) = 4\varepsilon \left\{ \left(\frac{\sigma}{r}\right)^{12} - \left(\frac{\sigma}{r}\right)^{6} \right\}. \tag{2.83}$$

In Eq. (2.83) the parameters σ and ε are the usual Lennard–Jones parameters; the values used in the sample calculation on helium, neon, and argon are given in Table V.

1. Cluster Thermodynamics

As the first illustration of the utility of path integral methods we examine calculations on the free energy of formation of argon clusters. The free energy of formation of clusters is an important quantity in the theory of nucleation kinetics,[41] and there has been a significant amount of work attempting to approximately determine such free energies since the early 1930s. Two important alternative approaches to path integral Monte Carlo for the calculation of free energies of formation are classical Monte Carlo[37,41] and the normal mode approach.[42] Classical Monte Carlo methods to study the free energy of argon clusters have been used by Lee et al.[37] and by Garcia and Torroja.[43] By ignoring quantum effects, classical Monte Carlo methods are best at high temperatures. To determine the free energies of clusters at lower temperatures, where the effects of quantum mechanics can be expected to be more important, a common approach has been to treat the clusters as large polyatomic mole-

TABLE VI
The Gibbs Free Energy of a 13-Particle Lennard–Jones Representation of an Argon Cluster
at 1 Atmosphere Pressure

T	$\Delta G_{cm}/k_B T$	$\Delta G_{nm}/k_B T$	$\Delta G_{qm}/k_B T$
10	-408.2 ± 0.3	-383.8	-383.6 ± 0.4
15	-244.4 ± 0.6	-220.3	-232.0 ± 0.6
20	-161.6 ± 0.5	-137.6	-158.2 ± 0.5
25	-117.4 ± 0.4	-87.7	-115.7 ± 0.4
30	-88.1 ± 0.5	-54.4	-86.9 ± 0.5

cules and incorporate the vibrational modes within the standard harmonic approximation.[42] The harmonic approximations are deficient to the extent that anharmonicities are important. Furthermore, in the harmonic approximation, it is assumed that only one structure is important to the determination of the thermodynamic properties. In actuality at finite temperatures there is no uniquely defined structure, and a variety of atomic configurations may be important in the proper description of the cluster. The harmonic approximation can be expected to work best as the temperature approaches 0 K. It is worth questioning whether there is a range of temperatures over which neither the harmonic approximation nor classical Monte Carlo methods are adequate. In Table VI we present the Gibbs free energy of formation of 13 atom argon clusters at 1 atmosphere pressure for a range of temperatures from 10 to 30 K. The detailed methods to determine Gibbs free energies using Monte Carlo methods are somewhat subtle and unimportant for our current purposes. The methods to determine the free energy are described elsewhere.[13] In Table VI the results of classical Monte Carlo, ΔG_{cm}, the normal mode method, ΔG_{nm}, and quantum path integral Monte Carlo ΔG_{qm} are given as a function of temperature. From the table it is evident that in the range of temperatures from about 15 to 20 K neither the normal mode method nor classical Monte Carlo provides a correct description of the system. In this region anharmonicities and quantum effects are simultaneously important.

2. Cluster Melting

As a second example of the utility of equilibrium path integral methods in examining the degree of quantum effects in many-body systems, we discuss recent work on "melting" in small clusters. This problem has been approached with classical methods by a number of investigators.[44–54] Recently, Beck et al.[55] examined the issue of the importance of quantum effects on the transition temperature. They computed the root mean square (rms) bond length fluctuations in the cluster as a function of temperature both classically and quantum mechanically. The authors also calculated distributions of po-

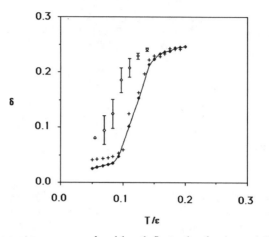

Figure 5. The root mean square bond length fluctuation for Ar_7 and Ne_7 clusters as a function of reduced temperature T/ε. The classical results for argon and neon are identical because of the law of corresponding states and are shown as the solid line. The crosses (open diamonds) are the quantum mechanical Ar_7 (Ne_7) results.

tential energies accessed along the Metropolis Monte Carlo walk to compare the coexistence behavior with and without the inclusion of quantum effects.

The results for argon and neon are given in Figure 5. The effects of quantum mechanics on the rms bond length fluctuations in argon are small, except at low temperature, where zero point motions are important. For neon, however, we see a significant quantum mechanical effect on the transition temperature. The transition temperature is lowered by approximately 30% in the quantum calculation relative to the classical calculation. Moreover, as discussed in Ref. 55, no coexistence of solid-like and liquid-like forms was observed in the quantum mechanical neon studies. This is in contrast to the classical results. We also see in Fig. 5 that the quantum mechanical rms bond fluctuations in neon are nearly 10% as the temperature approaches zero, a value that historically has been associated with melting in bulk systems.[56]

Recently, several experiments have been carried out on cluster systems to investigate the possibility of experimentally observing melting behavior in clusters.[57-60] These experiments have generally been performed on systems composed of a chromophoric species located in or on a cluster of argon atoms. Classical simulations on mixed argon–chromophore clusters are likely to yield useful results for comparison to experiment. However, calculations on mixed neon–chromophore systems will require the inclusion of quantum effects. Softening of the clusters and shifts in melting temperatures are in principle observable in spectroscopic experiments, and comparisons between classical and quantum studies, on the one hand, and experimental results, on the other, should provide an interesting testing ground for the examination of quantum effects on melting in small systems. Such quantum path integral studies of mixed systems are currently underway.

3. Fluids

An important capability of general numerical tools is that they can provide insight into complex problems that is otherwise difficult to obtain. For example, by providing numerical benchmarks for the properties of systems whose interactions are under control, such methods can greatly assist in the development of simpler, analytic models. The physical basis for such models can be identified and their numerical quality can be studied in an unambiguous way without the inevitable ambiguities related to the adequacy of empirical microscopic force laws that cloud comparisons with experimental data.

The development of the WCA theory[40] of dense classical fluids is an example of this type of use of numerical methods. As developed in detail elsewhere,[40] the WCA model divides the potential energy into attractive and repulsive portions and takes as a reference system the fluid structure determined by the repulsive portions of the interactions. The effects of the attractive portions of the interactions are then included through low-order perturbation techniques,

It is interesting to ask to what extent the classical WCA model can be extended to quantum mechanical systems. As a simple example of the use of the present methods to resolve such questions, we present in Fig. 6 classical and quantum mechanical radial distribution functions obtained for a 64-

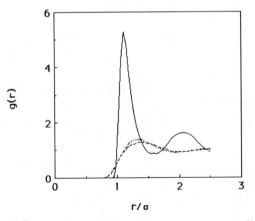

Figure 6.   The radial distribution function for a Lennard-Jones model of helium. The density is $\rho\sigma^3 = 0.365$ and the temperature is $T/\varepsilon = 0.5$. The dashed lines compare the quantum mechanical results obtained from a calculation versus the corresponding quantum mechanical results obtained with a WCA truncated potential. The corresponding classical result is shown for comparison.

particle model of liquid helium in a particular thermodynamic state. The classical results were obtained using a 64-particle, minimum image, spherical cutoff ($R_c = 2.5\sigma$) calculation[61] based on a Lennard–Jones interaction. Corresponding quantum mechanical calculations were performed for the same model interaction using the gradient partial averaging method described above and elsewhere.[28] For the particular thermodynamic state involved, these quantum mechanical results are fully converged with respect to the number of Fourier path variables included and can be regarded as effectively exact. Although the thermodynamic state involved is sufficiently high in temperature that particle statistics do not play a fundamental role, significant differences between the classical and quantum mechanical results in Fig. 6 indicate that quantum effects are nonetheless quite important. We also present in Fig. 6 results of similar quantum mechanical calculations performed on a WCA-truncated version of the Lennard–Jones potential. We see from the figure that this WCA-like calculation accurately reproduces the full quantum mechanical results. Although much more work would be required before one could form an informed opinion, this result would suggest that it might be possible to develop a quantum mechanical WCA model similar in nature to its classical counterpart. At present, there is no quantum mechanical analogue of the analytic, hard-core reference system solution that makes the classical WCA method so attractive. A *numerical* quantum mechanical WCA model is, nonetheless, potentially useful since calculations based on the truncated Lennard–Jones potential are substantially faster computationally than those based on the complete interaction.

III. DYNAMICS

As discussed in the previous section, significant progress has been made in the study of time-independent quantum mechanical systems. In particular, various Monte Carlo and numerical path integral methods are emerging as effective tools for the treatment of a wide range of problems involving both electronic[62-66] and heavy particle degrees of freedom.[3]

In this section we consider ongoing efforts to generalize these equilibrium methods to permit their application to problems in "real time" quantum dynamics. In principle, this generalization is straightforward. In practice, however, the practical realization of such lines of development have in the past been frustrated by a variety of technical issues. Although it is possible to phrase the construction of finite temperature time correlation functions in such a way that the problem becomes formally analogous to the equilibrium tasks discussed above, the appearance of complex, highly oscillatory exponential quantities as opposed to the simple, equilibrium Boltzmann-like factors greatly complicates the direct application of standard methods.

This section describes one approach to overcoming these intrinsic numerical difficulties based on the introduction of a formally exact, yet computationally practical analogue of the stationary phase method.[6,9,31,32,67-72] We intend the present discussion to be a convenient summary of information concerning these "stationary phase Monte Carlo" techniques and hope that it will serve to make these methods more accessible to the general dynamics community. We wish to emphasize that although preliminary results continue to be encouraging, these methods are currently under development and are thus at present incompletely characterized.

A. Correlation Function Preliminaries

In what follows we will consider two related types of time correlation functions. The "thermally symmetrized" correlation function of A and B is defined by

$$G_{AB}(t) = \frac{\text{tr}[Ae^{-\beta_c^* H} B e^{-\beta_c H}]}{\text{tr}[e^{-\beta H}]}, \tag{3.1}$$

where the complex temperature β_c is given by

$$\beta_c = \beta/2 + it/\hbar. \tag{3.2}$$

The general properties of these functions are discussed by Berne and Harp.[73] Following Miller et al.[74] a number of workers have utilized these thermally symmetrized correlation functions in the calculation of thermal rate coefficients. $G_{AB}(t)$ is related to the more familiar quantum mechanical time correlation function $C_{AB}(t)$,

$$C_{AB}(t) = \frac{\text{tr}[e^{-\beta H} A e^{iHt/\hbar} B e^{-iHt/\hbar}]}{\text{tr}[e^{-\beta H}]}, \tag{3.3}$$

through the Fourier transform identity

$$G_{AB}(\omega) = e^{-\beta\hbar\omega/2} C_{AB}(\omega). \tag{3.4}$$

In the time domain, this relationship becomes

$$C_{AB}(t) = G_{AB}(t - i\beta\hbar/2), \tag{3.5}$$

an exact extension of an approximation due to Schofield.[75-77] It is clear that both correlation functions contain the same physical information. Propagators in the thermally symmetrized form are always paired with corresponding thermal Boltzmann factors, a fact that may offer some computational advantage. As illustrated below, however, the present methods have been used successfully for the calculation of both types of time correlation functions.

B. Pseudo-Classical Formulation

The thermally symmetrized time correlation functions have an interesting mathematical structure that is useful in discussing the present problem. This structure helps clarify the connections between the present, less familiar developments and more traditional methods. In particular, we will find that it is possible to construct a *formally exact* procedure for the calculation of time correlation functions that strongly parallels familiar classical dynamical methods. Since the resulting formalism contains an essentially continuous spectrum of dynamical techniques, ranging from approximate classical and semiclassical approaches to the numerically exact, real time path integral methods, it serves as a convenient organizational device for the present discussion. This approach also serves to emphasize the separate roles of equilibrium and dynamical quantum mechanical effects in time correlation functions.

We begin with two, somewhat unrelated observations. First, we note that the classical probability of traveling from the point x at time zero to the point x' at time t in a system characterized by a temperature T, P_{cm}, is given by

$$P_{cm}(x \to x', T, t) = \int dp\, f(p, T)\delta(x' - x'(x, p, t)). \tag{3.6}$$

In this equation the integration is over all possible initial particle momenta p, $f(p, T)$ represents the equilibrium probability distribution function for the initial momenta in a system at a temperature T, and $x'(x, p, t)$ symbolizes the final position of a classical trajectory that began at point x with an initial momentum p and was propagated forward for a time t. For notational simplicity we hereafter suppress much of the functional dependence of the various quantities in Eq. (3.6) on secondary variables. Using the basic properties of the delta function in Eq. (3.6), it is easy to show that

$$P_{cm}(x \to x') = \sum_n \frac{f(p_n)}{|(\partial x'/\partial p)_n|}, \tag{3.7}$$

where the sum is over all roots p_n that satisfy

$$x'(x, p_n, t) = x'. \tag{3.8}$$

The classical probability is thus given by a sum of individual probabilities, each term arising from a classical path that connects the points x and x' in a physical time t. Since each trajectory is required to begin at a common point x, the associated statistical weight in Eq. (3.7) involves only the kinetic energy portion of the classical Boltzmann distribution. The inverse Jacobian structure of Eq. (3.7) admits the possible existence of "rainbow" singularities[78] in the classical transition probability.

Next, we give concrete form to the expression for the symmetrized correlation function, $G_{AB}(t)$. Evaluating the quantum mechanical traces in Eq. (3.1) in the coordinate representation gives (considering the case where the operators A and B are diagonal),

$$G_{AB}(t) = \frac{\int dx\,dx'\,|\langle x'|e^{-\beta_c H}|x\rangle|^2 A(x)B(x')}{\int dx\,dx'\,|\langle x'|e^{-\beta_c H}|x\rangle|^2}. \tag{3.9}$$

Equation (3.9) is reminiscent of the equilibrium thermodynamic averages of the previous section. Here, however, the variables x and x' are coupled *dynamically* through the appropriate *complex* temperature density matrix element.

Motivated by the form of Eq. (3.9) and by developments described below, we define a quantum mechanical object P_{qm} by the expression

$$P_{qm}(x \to x') = \frac{|\langle x'|\exp(-\beta_c H)|x\rangle|^2}{\langle x|\exp(-\beta H)|x\rangle}. \tag{3.10}$$

From Eq. (3.10) we see that P_{qm} has a number of useful properties. In particular, it is

1. Positive
2. Normalized with respect to integration over x'
3. Given in the semiclassical limit by[79]

$$\lim_{\hbar \to 0} P_{qm}(x \to x') = \left| \sum_n \frac{1}{(2\pi m k_B T)^{1/4}} \frac{\exp(-\beta_c p_n^2/2m)}{\sqrt{(\partial x'/\partial p)_n}} \right|^2, \tag{3.11}$$

where the sum is over all classical trajectories that connect x and x' in a time t (i.e., over the solutions of Eq. 3.8). In the purely classical limit (obtained by dropping the interference terms from Eq. 3.11), P_{qm} reduces to the classical transition probability given by Eq. (3.7). We therefore tentatively accept P_{qm} as a quantum mechanical generalization of the corresponding classical probability. P_{qm} has other properties that will further reinforce this interpretation.

Interference effects are inherent in P_{qm} since it is built from the absolute square of a quantum mechanical amplitude and not from the sum of probabilities.

Using Eq. (3.10) we can write the symmetrized time correlation function in a particularly convenient form. Multiplying and dividing the numerator of Eq. (3.9) by the density matrix element

$$\rho_{qm}(x) = \langle x | e^{-\beta H} | x \rangle, \tag{3.12}$$

and using the definition of P_{qm} (Eq. 3.10), $G_{AB}(t)$ becomes

$$G_{AB}(t) = \frac{\int dx \, \rho_{qm}(x) \int dx' \, P_{qm}(x \to x') A(x) B(x')}{\int dx \, \rho_{qm}(x)}. \tag{3.13}$$

The structure of Eq. (3.13) is that of a mixed equilibrium/dynamical average. The x average in Eq. (3.13) is over "initial" positions whose statistical weights are specified by the equilibrium quantum mechanical density $\rho_{qm}(x)$. For each specified initial position, the x' integration sums over all possible "final" positions, weighting each contribution by the "probability" of the $x \to x'$ transition.

The link between Eq. (3.13) and more traditional classical approaches becomes more evident if we make use of a general property of the classical transition probability. From Eq. (3.6) it is straightforward to show that for a general function $g(x')$,

$$\int dx' \, P_{cm}(x \to x') g(x') = \int dp \, f(p, T) g(x'(x, p, t)). \tag{3.14}$$

In other words, Eq. (3.14) is a general route for reexpressing final coordinate space averages as integrals over initial momenta. Using Eqs. (3.13) and (3.14), we thus see that the symmetrized quantum time correlation function can be written *exactly* as

$$G_{AB}(t) = \frac{\int dx \, dp \, \rho_{qm}(x) f(p, T) \{ A(x) B(x') P_{qm}(x \to x') / P_{cm}(x \to x') \}}{\int dx \, dp \, \rho_{qm}(x) f(p, T)}. \tag{3.15}$$

It is to be understood that x' in the above expression is the trajectory function $x'(x, p, t)$, and is generated by the same mechanics that underlies the classical transition probability P_{cm}.

The pseudoclassical form for $G_{AB}(t)$, Eq. (3.15), gives the *exact* quantum mechanical correlation function in a form that is strikingly similar to the customary classical expression. In the present notation, the corresponding classical result would be

$$G_{AB}^{cm}(t) = \frac{\int dx\, dp\, \rho_{cm}(x) f(p, T) \{A(x)B(x')\}}{\int dx\, dp\, \rho_{cm}(x) f(p, T)}. \tag{3.16}$$

In Eq. (3.15) modifications of the usual classical equilibrium and dynamical procedures produce the proper quantum mechanical result. On the equilibrium side, Eq. (3.15) uses a quantum mechanical rather than a purely classical distribution of initial positions, thus removing an obvious classical shortcoming. Even with this correct, short-time distribution, classical dynamics (in Eq. 3.15 a formal device for sweeping out all possible final positions) does not generally yield the correct probability distribution of final positions. Equation (3.15) corrects for this limitation by reweighting of the classically obtained final trajectory positions with a "dynamical correction factor," P_{qm}/P_{cm}. The dynamical correction factor P_{qm}/P_{cm} is available, in principle, through Eqs. (3.6) and (3.10). Although we expect from its mathematical definition and from its physical context that this object is, in general, a complicated one, its behavior in certain limiting situations can be anticipated. From Eqs. (3.7) and (3.11), we see that *semiclassically* P_{qm}/P_{cm} is given in terms of information associated with the *classical* trajectories that connect x and x' by

$$\lim_{\hbar \to 0} \frac{P_{qm}(x \to x')}{P_{cm}(x \to x')} = \frac{|\sum_n \exp(-\beta_c p_n^2/2m)/\sqrt{(\partial x'/\partial p)_n}|^2}{\sum_n \exp(-\beta p_n^2/2m)/|(\partial x'/\partial p)_n|}. \tag{3.17}$$

Quantum mechanical effects at this level enter through interference effects between various classical alternatives. In Miller's words,[79,80] we have in this limit "classical mechanics plus superposition." When only a single classical trajectory contributes to Eq. (3.17) [e.g., quadratic potentials, short time (cf. Eq. 3.33)], the semiclassical dynamical correction factor reduces to unity. In other words, within the semiclassical approximation the dominant quantum mechanical effect on the correlation function at short times is equilibrium in nature, the replacement of $\rho_{cm}(x)$ by its quantum mechanical counterpart. At longer times there will generally be many classical trajectories contributing to Eq. (3.17). These paths contribute linearly to the semiclassical *amplitude* in Eq. (3.17), but, because of the modulus structure of Eq. (3.17), no longer contribute individually to the *probability*. Interference effects between the various alternatives alter the simple classical dynamics. When there are many such paths, the details of this interference structure is potentially quite involved. If the specific property under study depends on the intricate details of this *entire* interference structure, an accurate description will require that we find a means for including the effects of all such paths. There may, however, be circumstances in which such complete detail is unnecessary. To suggest how such a circumstance might occur, we examine the nature of the general interference term in Eq. (3.11) or Eq. (3.17) arising from two paths, k and l. Explicitly multiplying out the absolute square in Eq. (3.11), we see that the coefficient of the interference term corresponding to paths k and l involves the Boltzmann-like factor, $\exp(-\beta(p_k^2 + p_l^2)/2)$. Thus only thermally accessible paths will contribute significantly to the final result. More subtly, this same analysis also shows that coefficient of the interference term arising from paths k and l contains a phase oscillation that is determined by the energy *difference* between the two trajectories. Specifically, the coefficient of the interference term is proportional to $\exp(-it(p_k^2 - p_l^2)/2\hbar)$. For large values of t, the resulting phase oscillations may be sufficiently severe that the interference term will not survive the average over final positions contained in Eq. (3.13). Finally, we note that in some circumstances the long time limit of the dynamical correction factor can be expressed in terms of purely equilibrium quantities. *If the system is dissipative*, (i.e., if the probability that the dynamics produces a particular final position at long times approaches the value given by the equilibrium density), then we have

$$\lim_{t \to \text{large}} \frac{P_{\text{qm}}(x \to x')}{P_{\text{cm}}(x \to x')} = \frac{\rho_{\text{qm}}(x')/Q_{\text{qm}}}{\rho_{\text{cm}}(x')/Q_{\text{cm}}}, \qquad (3.18)$$

where Q_{qm} and Q_{cm} are the quantum and classical partition functions, respectively. For nondissipatative systems, the transition probabilities on the left-hand side of Eq. (3.18) do not approach a time-independent value and Eq. (3.18) is thus not appropriate.

C. Monte Carlo Formulation of the Dynamical Problem

We now turn to the task of developing a practical means for computing the quantum mechanical time correlation functions. In the present formulation, this will ultimately require either the calculation of complex temperature density matrix elements, or, more generally, averages over such matrix elements of the type indicated in Eq. (3.6). Such problems are of the generic form of a high dimensional average of an integrand that becomes progressively more oscillatory as the physical time increases. These oscillations limit the utility of standard numerical approaches to relatively short physical times.

It is convenient to begin by noting that the path integral formalism of Section II readily generalizes to complex temperatures. In particular, it is not difficult to show[28,81] that the complex temperature density matrix element, $\langle x' | \exp(-\beta_c H) | x \rangle$, can be written as

$$\langle x' | e^{-\beta_c H} | x \rangle = \rho_{\text{fp}}(x, x'; \beta_c) \frac{\int d\mathbf{a} \exp(-\sum_{k=1}^{\infty} a_k^2/2\sigma_k^2 - \beta_c \langle V \rangle_{\mathbf{a}})}{\int d\mathbf{a} \exp(-\sum_{k=1}^{\infty} a_k^2/2\sigma_k^2)}. \qquad (3.19)$$

Here ρ_{fp} corresponds to the free-particle density matrix element evaluated at the complex temperature β_c,

$$\rho_{fp}(x, x'; \beta_c) = \left(\frac{m}{2\pi\hbar^2\beta_c}\right)^{1/2} \exp\left[\frac{-m(x' - x)^2}{2\hbar^2\beta_c}\right]. \tag{3.20}$$

The gaussian widths are given by

$$\sigma_k^2 = 2\beta_c\hbar^2/(m\pi^2k^2). \tag{3.21}$$

The quantum mechanical paths connecting x and x' are parameterized in terms of the Fourier coefficients $\mathbf{a}$ by

$$x_{\mathbf{a}}(u) = x + (x' - x)u + \sum_{k=1}^{\infty} a_k \sin(k\pi u), \tag{3.22}$$

where the dimensionless time variable u here ranges from zero to one. In terms of this reduced time variable the average along the path specified by Eq. (3.22) is given by

$$\langle V\rangle_{\mathbf{a}} = \int_0^1 du\, V(x(u)). \tag{3.23}$$

From Eq. (3.21) we see that the natural length scales in a dynamical problem are dictated by $|\beta_c|$. Consequently, at either low temperatures or at long times many auxiliary degrees of freedom will be required in order to obtain convergence with respect to the number of Fourier coefficients in Eq. (3.22).

Substituting the complex temperature density matrix elements into Eq. (3.9), we ultimately find that the thermally symmetrized time correlation function can be written as

$$G_{AB}(t) = \frac{\int dx\, dx'\, d\mathbf{a}\, d\mathbf{b}\, \rho([\mathbf{a}], [\mathbf{b}]) \exp(i\tau f([\mathbf{a}], [\mathbf{b}])) A(x) B(x')}{\int dx\, dx'\, d\mathbf{a}\, d\mathbf{b}\, \rho([\mathbf{a}], [\mathbf{b}]) \exp(i\tau f([\mathbf{a}], [\mathbf{b}]))}, \tag{3.24}$$

where τ is the ratio of the physical to thermal times $[\tau = t/(\beta\hbar/2)]$. By analogy with Eq. (3.22), the Fourier coefficients $\mathbf{b}$ parameterize the paths for the complex conjugate density matrix element in Eq. (3.6) that travel from x' back to x. Specifically,

$$x_{\mathbf{b}}(u) = x' + (x - x')u + \sum_{k=1}^{\infty} b_k \sin(k\pi u). \tag{3.25}$$

The weight and phase function in Eq. (3.24) depend on the paths $x_a(u)$ and $x_b(u)$ and are given by

$$\rho([a], [b]) = |\rho_{fp}(x, x'; \beta_c)|^2 e^{-[S_+([a]) + S_+([b])]}, \tag{3.26}$$

and

$$f([a], [b]) = S_-([a]) - S_-([b]). \tag{3.27}$$

We have suppressed much of the explicit dependence of the weight and phase functions on secondary variables in the interests of notational simplicity. The $S_\pm$ terms are given for the case of the a variables by

$$S_\pm([a]) = \sum_{k=1}^{\infty} a_k^2/2s_k^2 \pm \frac{\beta}{2}\langle V \rangle_a, \tag{3.28}$$

where

$$s_k^2 = 2|\beta_c|^2 \hbar^2/(m\pi^2 k^2 (\beta/2)) \tag{3.29}$$

and where $\langle V \rangle_a$ is the average of the potential energy along the path $x_a(u)$, Eq. (3.23). The corresponding expressions for $S_\pm([b])$ terms are the obvious extensions of the results in Eq. (3.29). The action that appears in the complex temperature density matrix element expression can be written in terms of $S_\pm$ as

$$S(a) = \sum_{k=1}^{\infty} a_k^2/2\sigma_k^2 + \beta_c \langle V \rangle_a = S_+(a) - i\tau S_-(a). \tag{3.30}$$

Insight into the analytic structure of the symmetrized time correlation function and Eq. (3.24) can be gained by examining those paths that extremize the complex action, Eq. (3.30). As shown elsewhere,[82] the Fourier coefficients describing such paths (i.e., the coefficients that satisfy the requirement that $\partial S/\partial a_k = 0$) are the fixed points of the set of complex maps

$$a_k = \frac{\beta_c^2 \hbar^2}{m\pi^2 k^2} F_k(a), \tag{3.31}$$

where $F_k(a)$ is the kth Fourier sine component of the force along the path specified by the coefficients $\{a_k\}$,

$$F_k(a) = 2 \int_0^1 du \sin(k\pi u) \left\{ -\frac{dV(x(u))}{dx(u)} \right\}. \tag{3.32}$$

Solution(s) to this set of nonlinear equations determine the stationary paths for Eq. (3.24). In the limit of either small or large times, Eq. (3.31) becomes (cf. Eq. 3.2)

$$
a_k = \begin{cases} \dfrac{\{\beta/2\}^2\hbar^2}{m\pi^2k^2}F_k(\mathbf{a}) & t\text{ small} \\[4mm] -\dfrac{t^2}{m\pi^2k^2}F_k(\mathbf{a}) & t\text{ large.} \end{cases} \tag{3.33}
$$

From Eq. (3.28) it is easy to show that the short and long time limits in Eq. (3.33) are equivalent to the requirements that $\partial S_+(\mathbf{a})/\partial a_k = 0$ and $\partial S_-(\mathbf{a})/\partial a_k = 0$, respectively. In general, the significance of a particular path in Eq. (3.24) is governed by an interplay between equilibrium *and* dynamical considerations. The weight function ρ in Eq. (3.24) is essentially an equilibrium Boltzmann factor involving the combination of kinetic *plus* potential energies familiar from time-independent path integral applications. The phase function, on the other hand, is associated with the dynamical, kinetic *minus* potential energy combination. For short times, Eq. (3.33) indicates that equilibrium considerations dominate and the paths contributing most strongly to Eq. (3.24) are those that maximize ρ (and minimize the S_+ action). For longer times, however, the phase oscillations of the complex exponential become more severe, and the physically relevant paths shift from being those that *minimize* S_+ to those that *extremize* S_-. That is, the integrals in Eq. (3.24) shift with increasing time from steepest descent to stationary phase in character. In both limits, the Fourier coefficients that fix these paths (i.e., solve Eq. 3.31) approach the *real* axis.

It is interesting to note that Planck's constant disappears from the long time limit in Eq. (3.33), implying that the result is classical in nature. Indeed, Eq. (3.33) in this limit can be obtained by purely classical means by substituting Eq. (3.22) directly into Newton's equations of motion. In this limit Eq. (3.33) thus determines the "double-ended" classical trajectories, trajectories that start at the point x at time zero and travel to the prescribed point x' in a specified time t. As discussed elsewhere,[69] Eq. (3.33) and simulated annealing methods are a convenient route for the determination of these double-ended classical trajectories. In the present application, knowledge of their number and location is useful for characterizing the complexity of the integrals in Eq. (3.24) and in developing efficient Monte Carlo sampling strategies for its evaluation. Beyond this particular application, these double-ended trajectories are important in a number of other contexts. For example, as discussed by Pratt,[83] such trajectories can be used to characterize transition states in complex systems. Methods for locating stationary phase regions in functional integrals and classical trajectories satisfying double-ended boundary conditions have been discussed elsewhere.[69, 84] The interested reader is referred to the original literature for a discussion of the details.

D.   The Stationary Phase Monte Carlo Method

The discussion given above illustrates that the calculation of the thermally symmetrized time correlation functions can be reduced to the problem of estimating high dimensional averages of the type in Eq. (3.24). We examine this problem in prototype form by considering the integral

$$I(t) = \int dx\, \rho(x) e^{itf(x)}, \qquad (3.34)$$

where $\rho(x)$ is a positive weight function, $f(x)$ is a specified phase function, and t is a parameter we will ultimately associate with the physical time. Equation (3.34) will serve as a useful model for purposes of the following discussion and will simplify the presentation of the relevant ideas. We note, however, that our ultimate objective, Eq. (3.24), contains the additional feature that the weight and phase functions also depend on time. This additional time dependence prevents the direct use of what would otherwise be an obvious series of simplifications. For example, if our problem were simply of the form given in Eq. (3.34), then approaches familiar from related work on Radon transformations[85] would suggest that we rewrite $I(t)$ as

$$I(t) = \int d\omega\, e^{i\omega t} \int dx\, \rho(x)\delta(\omega - f(x)). \qquad (3.35)$$

This would compress the many-dimensional, complex integrations present in the original problem into a single, one-dimensional Fourier transform of an equilibrium-like average. Moreover, if the problem were purely as indicated in Eq. (3.34), and if we were only interested in the Fourier transform of $I(t)$, then we see from Eq. (3.35) that this could be written as

$$\hat{I}(\omega) = 2\pi \int dx\, \rho(x)\delta(\omega - f(x)). \qquad (3.36)$$

In this case the equilibrium methods described in the previous section would allow us to compute the frequency space transform *directly*. To date, this direct approach has not proved of utility for the calculation of time correlation functions, because of the additional time dependence in the density and phase functions.

An approach that has proved useful for the present problem begins with a physical observation. As the value of the parameter t in Eq. (3.34) increases, oscillations in the integrand produced by variations in the coordinate x typically become increasingly severe. If these oscillations become rapid on a length scale that is small relative to the natural length scale of the weight function $\rho(x)$, they effectively destroy any contribution to the final result arising from that region of the integrand. This argument, familiar from ordinary stationary phase developments,[86] indicates that $I(t)$ is dominated in the limit of large values of t by the behavior of the integrand in those regions where the phase function is locally constant. Phrased differently, in this limit the spatially local phase interference structure in the problem is irrelevant, and it is necessary to concentrate only the larger scale interference structure between the different stationary phase regions. In conventional approaches these observations are used to develop a computational algorithm based on locating the stationary phase points of the relevant phase function. Once located, information concerning the behavior of the integrand in these regions is assembled into an asymptotically valid approximation to the final result. The conventional stationary phase estimate[86] for Eq. (3.34) is thus

$$\lim_{t \to \infty} I(t) = \sum_n \rho(x_n) \exp(itf(x_n)) \left\{ \frac{2\pi}{-itf''(x_n)} \right\}^{1/2}, \qquad (3.37)$$

where the sum in Eq. (3.37) is over the roots of the equation $f'(x_n) = 0$.

The application of ordinary stationary phase methods to the calculation of time correlation functions is not our objective. Although frequently useful, these asymptotic methods are inherently approximate, contain no *a priori* measure of their absolute accuracy, and provide us with no practical route for systematic refinement. Furthermore, the conventional application of these methods is itself a difficult task, typically requiring the numerical solution of large dimensional, nonlinear equations to locate the required stationary phase points.

Although not our objective, these asymptotic methods do suggest a related and productive line of development. Comparing Eqs. (3.37) and (3.34), we see that applying the stationary phase method is tantamount to replacing the integrand in Eq. (3.34) according to the rule

$$\rho(x)e^{itf(x)} \to D_{sp}(x)\{\rho(x)e^{itf(x)}\}, \qquad (3.38)$$

where the action of $D_{sp}(x)$ on the original integrand is given in terms of the stationary phase quantities by

$$D_{sp}(x)\{\rho(x)\exp(itf(x))\} = \sum_n \delta(x - x_n) \left\{ \frac{2\pi}{-itf''(x_n)} \right\}^{1/2} \rho(x)\exp(itf(x)). \quad (3.39)$$

Physically unimportant but mathematically troublesome phase oscillations are thus removed by the action of the "stationary phase filter" or "damping function", $D_{sp}(x)$. Even though the original and modified integrands are qualitatively different, integrals over them are nonetheless asymptotically equivalent.

As a generalization of this idea, we search for functions $D(x)$ that leave the integral Eq. (3.34) invariant under the transformation

$$\rho(x)e^{itf(x)} \rightarrow D(x)\rho(x)e^{itf(x)}. \tag{3.40}$$

That is, we are looking for the formally exact analogues of the corresponding stationary phase result, Eq. (3.38). If we can find such functions, our remaining task will then be to use them to effect a practical simplification of Eq. (3.34). As indicated below, both steps of this program can be realized.

We obtain a formal solution to our problem if we define $D(x)$ by

$$D(x) = \int dy\, P(y)\frac{\rho(x-y)}{\rho(x)}e^{it\{f(x-y)-f(x)\}}, \tag{3.41}$$

where $P(y)$ is an arbitrary (normalized) probability distribution. By direct evaluation it is easy to show that Eq. (3.34) and

$$I(t) = \int dx\, D(x)\rho(x)e^{itf(x)} \tag{3.42}$$

are formally equivalent, provided that the x integration extends either over an infinite interval or over an interval on which the phase function is periodic. Equation (3.41) thus constitutes the first step in our proposed generalization.

Having found a formal expression for our damping function, Eq. (3.41), we are now free to exploit the arbitrariness of the choice of the probability distribution $P(y)$ to secure a practical simplification of the original numerical problem. Qualitatively, we wish to suppress the irrelevant phase oscillations in Eq. (3.34) by making $D(x)$ function as a "band pass" filter that is "transparent" in regions where the phase function is slowly varying and "opaque" elsewhere. We can meet this goal if we take $P(y)$ to be a prelimit delta function of length scale ε, $\delta_\varepsilon(y)$. With this choice of $P(y)$, the damping function is given by

$$D_\varepsilon(x) = \int dy\, \delta_\varepsilon(y)\frac{\rho(x-y)}{\rho(x)}e^{it\{f(x-y)-f(x)\}}. \tag{3.43}$$

The basic structure of $D_\varepsilon(x)$ can be seen if we evaluate Eq. (3.43) in the limit of large t. We find asymptotically (compare with Eqs. 3.38 and 3.39)

$$\lim_{t \to \infty} D_\varepsilon(x)\{\rho(x)\exp(itf(x))\} = \sum_n \delta_\varepsilon(x - x_n)\left\{\frac{2\pi}{-itf''(x_n)}\right\}^{1/2} \rho(x_n)\exp(itf(x_n)).$$

$$(3.44)$$

We emphasize that the asymptotic analysis leading to Eq. (3.44) has been performed solely to highlight a particular, formal property of $D_\varepsilon(x)$. It would obviously be illogical to reject stationary phase methods for the original problem, Eq. (3.34), and accept them for the related problem of evaluating the damping function, Eq. (3.43).

We now examine various practical strategies for calculating the damping function, $D_\varepsilon(x)$. It is convenient to define approximate damping functions by means of the expression

$$D_{0,\varepsilon}(x) = \int dy\, \delta_\varepsilon(y)e^{it\Delta f_0(x,y)}, \qquad (3.45)$$

where $\Delta f_0(x, y)$ is some approximation to the true phase difference $\Delta f(x, y)$, defined by

$$\Delta f(x, y) = f(x - y) - f(x). \qquad (3.46)$$

The choice of $\Delta f_0(x, y)$ will ultimately be a compromise between the conflicting goals of simplicity and accuracy. For any particular choice of $\Delta f_0(x, y)$, the difference between the exact and approximate damping functions can be written in integral form using Eqs. (3.43), (3.45), and (3.46) as

$$D(x) - D_{0,\varepsilon}(x) = \int dy\, \delta_\varepsilon(y)\left\{\frac{\rho(x - y)}{\rho(x)}e^{it\Delta f(x,y)} - e^{it\Delta f_0(x,y)}\right\}. \qquad (3.47)$$

Equation (3.47) is a convenient starting point for the development of an unbiased Monte Carlo estimator for "corrections" to these approximate damping functions. An important special case of these general expressions is the particular choice of $\delta_\varepsilon(y)$ to be a gaussian of width ε and $\Delta f_0(x, y)$ to be the second-order gradient expansion of $\Delta f(x, y)$,

$$\Delta f_0(x, y) = -f'(x)y + f''(x)y^2/2. \qquad (3.48)$$

With these choices, Eq. (3.47) produces the approximate "second-order" damping function

$$D_{0,\varepsilon}^{(2)}(x) = \frac{1}{\sqrt{1 - it\varepsilon^2 f''}}\exp[-(\varepsilon t f')^2/2(1 - it\varepsilon^2 f'')). \qquad (3.49)$$

Retaining only the first term in the gradient expansion of $\Delta f(x, y)$ yields the corresponding "first-order" approximation

$$D_{0,\varepsilon}^{(1)}(x) = \exp\left(-\frac{1}{2}(\varepsilon t f')^2\right). \tag{3.50}$$

As expected, at both levels of approximation the resulting damping function acts to suppress phase oscillations where $f'(x) \neq 0$ (i.e., in the nonstationary phase regions).

An important practical issue in this approach is the choice of the damping function width parameter ε. The choice of its parameter dictates the sharpness of the damping function. For small values of ε the width of the damping function tends to vary (cf. Eq. 3.49) as $1/\varepsilon^2$, and thus *decreases* with increasing values of ε. However, we see from Eq. (3.49) that for larger parameter values the width of the damping function *increases* linearly with ε. This analysis, developed in greater detail elsewhere,[67] suggests that there will be an optimal (real) range of ε values that will produce the maximal compression of the damping function. The qualitative reason for this ε-dependence can be understood by noting that applying the damping function to the integrand is, effectively, a "coarse-graining" procedure. Optimal filtering will occur if the length scale of this coarse-graining procedure is roughly of the same length scale as the width of the stationary phase region itself. Extension of these arguments to complex values of ε are described elsewhere.[67]

In actual applications of the present method, we will be required to evaluate the damping function $D_\varepsilon(x)$ for arbitrary values of x. One convenient approach is to write $D_\varepsilon(x)$ as

$$D_\varepsilon(x) = D_{0,\varepsilon}(x) + \{D_\varepsilon(x) - D_{0,\varepsilon}(x)\}_{\text{MC}}. \tag{3.51}$$

In Eq. (3.51) $D_{0,\varepsilon}(x)$ is an approximate damping function, typically the first-order gradient form in Eq. (3.40). In practice, if $D_{0,\varepsilon}(x)$ (or its modulus) is greater than a preselected threshold δ, an unbiased MC estimate of the correction to $D_{0,\varepsilon}(x)$, the term in braces in Eq. (3.51), is added to the result. Such an unbiased MC estimate of the correction to $D_{0,\varepsilon}(x)$ can be based on Eq. (3.47). If, on the other hand, $D_{0,\varepsilon}(x)$ is small (i.e., smaller than our preselected threshold), we assume in Eq. (3.51) that the x region in question is unimportant and that further corrections are unnecessary. In actual applications the threshold value δ is varied to assure that the results obtained are insensitive to the particular value chosen. This procedure is similar in spirit to commonly utilized approximation schemes in quantum chemical electronic structure calculations. In such applications the accuracy with which various matrix elements are computed is decided on the basis of a preliminary, approximate estimate of the magnitude of the term in question. Matrix elements deemed "significant" are evaluated accurately, while those that are less important are treated more approximately.

We now describe two possible uses of the damping function in the calculation of averages of the type appearing in Eq. (3.42). One strategy is to utilize $D_\varepsilon(x)$ as a variance reduction device in an otherwise conventional Monte Carlo procedure. In this "ρ-sampling" method we evaluate Eq. (3.42) using the points $\{x_n\}$, selected at random from the distribution $\rho(x)$, giving

$$I(t) = \frac{1}{N} \sum_n D(x_n) \exp(itf(x_n)). \tag{3.52}$$

Using the same quadrature points the conventional Monte Carlo result based on Eq. (3.34) would be

$$I(t) = \frac{1}{N} \sum_n \exp(itf(x_n)). \tag{3.53}$$

As documented elsewhere, the damping function, by performing a prior synthesis of the integrand's local phase interference structure, can accelerate the convergence of Eq. (3.52) relative to Eq. (3.53). Provided that the ρ distribution adequately covers the physically relevant stationary phase regions and that we can actually compute $D_\varepsilon(x)$ for arbitrary values of x, this approach offers a possible method for estimating the averages of interest. Alternatively, we can use the damping function to devise improved importance sampling procedures that from the outset build in the physically relevant regions rather than simply using the function to retroactively kill off contributions arising from unwisely chosen quadrature points. To do this it is useful to incorporate a positive approximation (e.g., Eq. 3.50) $D_0(x)$ for the damping function into an importance sampling scheme by writing Eq. (3.42) as

$$I(t) = \int dx \, \rho(x) D_0(x) \left[\frac{D(x)}{D_0(x)} \right] e^{itf(x)}. \tag{3.54}$$

This can be thought of as an average over the modified weight function $\rho(x)D_0(x)$ rather than over the original weight $\rho(x)$. Numerically, we can thus evaluate Eq. (3.54) using $\rho(x)D_0(x)$ as the basis of a Monte Carlo importance sampling procedure. If $D_0(x)$ is chosen wisely, quadrature points selected by this "$\rho-D_0$" importance sampling procedure will better reflect the underlying stationary phase structure of Eq. (3.34). In situations where multiple stationary phase regions exist, care must be exercised to assure that all such regions are properly sampled. We are aided in this matter by the analytic characterization of these regions as extrema of the phase function. Furthermore, guidance concerning the adequacy of particular MC sampling strategies can be gained by comparing results obtained independently from ρ and $\rho-D_0$ importance sampling approaches. The calculation of the normalization integral, the average of $D_0(x)$ over the weight $\rho(x)$, poses no special difficulties.

To illustrate the stationary phase Monte Carlo method, we use the ρ-sampling approach to evaluate the one-dimensional, prototype integral

$$I(t) = \frac{\int_{-\infty}^{\infty} dx\, e^{-x^2/2} e^{it(x^2-1)^2}}{\int_{-\infty}^{\infty} dx\, e^{-x^2/2}}.$$

(3.55)

This integral is of the generic form of Eq. (3.34), with ρ being a simple gaussian of unit width and $f(x) = (x^2 - 1)^2$. As illustrated in Fig. 7, the complex

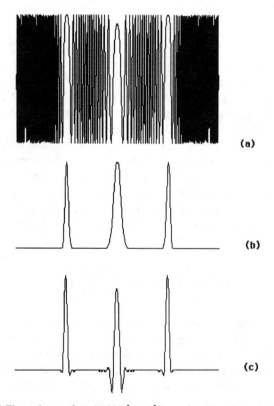

(a)

(b)

(c)

Figure 7.   (a) The real part of $\exp(i\,100(x^2-1)^2)$ as a function of x over the range $(-2, 2)$. b) The first order-damping function $D_{0,\varepsilon}(x)$ for this problem, Eq. (3.50), for $\varepsilon = 0.04$. (c) The real part of $D_{0,\varepsilon}(x)\exp(i\,100(x^2-1)^2)$ as a function of x over the range $(-2, 2)$.

exponential in the integrand is oscillatory, except in the vicinity of the stationary phase points $x = 0$ and $x = \pm 1$. In a direct Monte Carlo evaluation of Eq. (3.55) (Eq. 3.53), we would have to reproduce numerically the details of these phase oscillations to obtain a reliable estimate of $I(t)$. In the stationary phase Monte Carlo approach, Eq. (3.52), most of these phase oscillations are removed by the action of the damping function. The efficiency of the present approach is a balance between the savings introduced by the damping function's removal of these troublesome phase oscillations versus the additional numerical cost of constructing the damping function itself. Figure 8 shows plots of the modulus of the damping function near $x = -1$ and $x = 0$

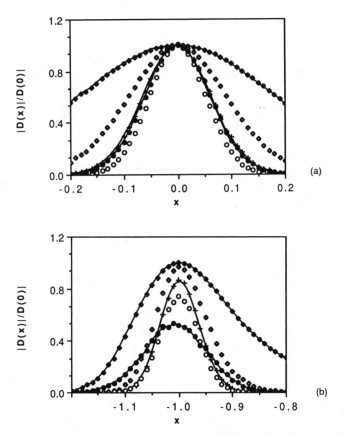

Figure 8. (a) Plots of $|D_\varepsilon(x)|/|D_\varepsilon(x = 0)|$ in the vicinity of $x = 0$ as a function of ε for the quartic example of Section III ($t = 100$). Plots correspond to $\varepsilon = 0.01$ (solid diamonds), 0.02 (open diamond), 0.03 (cross), 0.04 (open circle), and 0.08 (solid circle); (b) as in a, except that plot covers the vicinity of $x = -1$.

TABLE VII

Real and Imaginary Values for the Quartic Model Integral of Section III (Eq. 3.55) Computed for Various Values of t and ε[a]

	SPMC-D$_0$				SPMC-Corrected			
ε	Re($I(t)$)	σ	Im($I(t)$)	σ	Re($I(t)$)	σ	Im($I(t)$)	σ
				$t = 100$				
0	0.420(−1)	0.12(−2)	−0.171(−1)	0.13(−2)				
0.010	0.461(−1)	0.09(−2)	−0.187(−1)	0.07(−2)	0.429(−1)	0.09(−2)	−0.169(−1)	0.09(−2)
0.020	0.505(−1)	0.07(−2)	−0.225(−1)	0.06(−2)	0.428(−1)	0.05(−2)	−0.179(−1)	0.06(−2)
0.030	0.545(−1)	0.06(−2)	−0.268(−1)	0.04(−2)	0.428(−1)	0.06(−2)	−0.177(−1)	0.05(−2)
0.040	0.552(−1)	0.05(−2)	−0.282(−1)	0.03(−2)	0.423(−1)	0.06(−2)	−0.186(−1)	0.06(−2)
				$t = 10{,}000$				
0	−0.105(−2)	0.116(−2)	0.264(−2)	0.132(−2)				
0.001	−0.131(−2)	0.032(−2)	0.490(−2)	0.015(−2)	−0.146(−2)	0.033(−2)	0.522(−2)	0.027(−2)
0.002	−0.109(−2)	0.017(−2)	0.446(−2)	0.013(−2)	−0.161(−2)	0.023(−2)	0.524(−2)	0.019(−2)
0.003	−0.096(−2)	0.016(−2)	0.324(−2)	0.009(−2)	−0.140(−2)	0.025(−2)	0.539(−2)	0.012(−2)
0.004	−0.145(−2)	0.018(−2)	0.196(−2)	0.007(−2)	−0.124(−2)	0.028(−2)	0.519(−2)	0.020(−2)

[a] For comparison, we list stationary phase Monte Carlo values (SPMC) for both corrected and uncorrected first-order gradient approximations to the damping function (cf. Eqs. 3.47 and 3.50). As discussed in the text, corrections were obtained by adding an unbiased, one point estimate of the difference between the first-order gradient approximation and the exact damping function to the uncorrected values when the magnitude of the uncorrected results exceeded a preselected threshold (10^{-6} for the present calculations). For each individual t value all methods utilized the same number of quadrature points (200,000). The relative efficiencies of the conventional and SPMC methods can thus be judged by comparing the squares of their corresponding statistical errors. As a measure of the absolute accuracy of the present results, brute force Monte Carlo estimates of the integral using 2×10^8 points gave $(0.4275 \pm 0.0005, -0.1792 \pm 0.0005) \times 10^{-1}$ at $t = 100$ and $(-0.1483 \pm 0.0050, 0.5299 \pm 0.0050) \times 10^{-2}$ at $t = 10{,}000$ while the corresponding conventional stationary phase method gives $(0.429, -0.181) \times 10^{-1}$ and $(-0.141, 0.532) \times 10^{-2}$ at $t = 100$ and $t = 10{,}000$, respectively.

computed numerically from Eq. (3.43) for various values of the width parameter ε. As expected from the discussion given above, the width of the damping function varies with ε, first decreasing and ultimately increasing as ε is increased from a value of zero. Intuitively, we expect efficiency of the present approach will be greatest when the width of damping function is adjusted to its minimum value. With ε in this optimal range, the damping function has isolated the individual stationary phase regions as much as possible by maximally suppressing the integrand's phase oscillations. From Fig. 8 we see that this "optimal" value of ε corresponds to approximately $\varepsilon = 0.04$ for $t = 100$. From Eqs. (3.50) and (3.55) the first-order gradient approximation to the damping function is given by

$$D_{0,\varepsilon}^{(1)}(x) = \exp(-2\varepsilon^2 t^2 x^2 (x^2 - 1)^2). \tag{3.56}$$

This approximate result is plotted in Fig. 7b for $t = 100$ and $\varepsilon = 0.04$. Comparing Fig. 7b and Fig. 8, we see that the first-order gradient result is a qualitatively correct representation of the modulus of the exact damping function. This conclusion is borne out in practice, and there proves to be no special difficulties in using Eq. (3.51) to estimate the differences between the first-order and exact damping functions. The effect of applying this first-order damping function to the complex exponential in the present problem is shown in Fig. 7c. Table VII lists values of $I(t)$ obtained using the ρ-sampling approach with various values of ε. For comparison, we have tabulated both the first-order gradient and exact results. The savings produced by the stationary phase Monte Carlo approach in Table VII relative to ordinary Monte Carlo methods is significant, amounting to approximately a factor of 40 for $t = 10,000$. Based on previous experience, we would expect similar results for $\rho-D_0$ sampling methods if applied to integrals of the present type.

E. The Stationary Phase Monte Carlo Calculation of Thermally Symmetrized Time Correlation Functions

We now consider the application of the stationary phase Monte Carlo method to the problem of calculating the thermally symmetrized time correlation function $G_{AB}(t)$. Using Eqs. (3.24) and (3.43), we have for the damping function

$$D_\varepsilon([\mathbf{a}], [\mathbf{b}]) = \int d\mathbf{a}'\, d\mathbf{b}'\, \delta_\varepsilon(\mathbf{a}', \mathbf{b}') \frac{\rho([\mathbf{a} - \mathbf{a}'], [\mathbf{b} - \mathbf{b}'])}{\rho([\mathbf{a}], [\mathbf{b}])}$$
$$\times \exp(i\tau\{f([\mathbf{a} - \mathbf{a}'], [\mathbf{b} - \mathbf{b}']) - f([\mathbf{a}], [\mathbf{b}])\}). \tag{3.57}$$

Inserting this damping function into *both* the numerator and denominator of Eq. (3.24) produces

$$G_{AB}(t) = \frac{\int dx\,dx'\,da\,db\,\rho([a],[b])D_\varepsilon([a],[b])\exp(i\tau f([a],[b]))A(x)B(x')}{\int dx\,dx'\,da\,db\,\rho([a],[b])D_\varepsilon([a],[b])\exp(i\tau f([a],[b]))}.$$

(3.58)

Equation (3.58) is the basic result of the present section. It illustrates that the thermally symmetrized time correlation function can be written as the ratio of two terms, each one of which is of the generic form of Eq. (3.34). We can thus apply the stationary phase Monte Carlo techniques developed above separately to the calculation of the numerator and denominator of Eq. (3.58). It is important to recognize that the same damping function appears in both numerator and denominator of Eq. (3.58), a consequence of including only convolutions over **a** and **b** degrees of freedom in Eq. (3.57). We have found no difficulties to date caused by not damping the x and x' degrees of freedom. We know on the basis of general arguments presented above that the damping function leaves the numerator and denominator individually unchanged while simplifying the troublesome phase oscillations otherwise present. We also know that if the damping function is computed exactly, the results obtained from Eq. (3.58) are formally independent of the choice of the probability distribution function that defines the damping function. In the applications discussed here, we have found the first-order gradient result (cf. Eq. 3.50)

$$D_{0,\varepsilon}([a],[b]) = \exp\left(-\frac{1}{2}\sum_{k=1}^{\infty} \tau^2\varepsilon_k^2\{(\partial f([a],[b])/\partial a_k)^2 + (\partial f([a],[b])/\partial b_k)^2\}\right)$$

(3.59)

to be especially convenient. This expression involves only first derivatives of the phase function. If required, corrections to Eq. (3.59) can be computed by a Monte Carlo procedure based on the multidimensional generalizations of Eqs. (3.47) and (3.51). Using the approximation of replacing the full damping function by Eq. (3.59), the symmetrized time correlation function becomes

$$G_{AB}(t) \approx \frac{\langle\exp(i\tau f([a],[b]))A(x)B(x')\rangle}{\langle\exp(i\tau f([a],[b]))\rangle},$$

(3.60)

where the brackets imply averages over $\rho D_{0,\varepsilon}$.

1. Application of Stationary Phase Monte Carlo Methods to the Calculation of Thermally Symmetrized Time Correlation Functions

We now present a simple application of the stationary phase Monte Carlo (SPMC)-approach to the calculation of thermally symmetrized time correla-

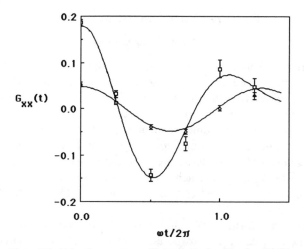

Figure 9.   Plots of $G_{xx}(t)$ for the quartic oscillator example of Section III. The units for $G_{xx}(t)$ are atomic units. Times shown are in multiples $2\pi/\omega$, where ω is the period of a simple oscillator whose force constant and mass are numerically equal to the corresponding parameters for the present quartic problem. The two curves correspond to results for two different temperatures of $\beta\hbar\omega = 1.0$ and 4.0.

tion functions. Other examples that document the feasibility of the basic approach have been considered elsewhere.[68] Figure 9 shows the dipole auto-correlation function $G_{xx}(t)$, computed for the quartic oscillator $V(x) = kx^4$. The mass was taken to be that of a proton and the proportionality constant k was chosen to be 9.14×10^{-3} au. For reference, if the potential were quadratic rather than quartic and if the same mass and force constant were used, the effective oscillator frequency would be $\hbar\omega/k_B = 1000$ K. The quartic oscillator is intrinsically anharmonic, as can be seen from the temperature dependence of the results in Fig. 9. At low temperatures, only transitions from the ground to first excited state are significant and thus the behavior of $G_{xx}(t)$ is effectively harmonic. The period predicted on the basis of the energy spacing of the ground and excited state of the quartic oscillator is approximately 2600 au, in good agreement with the low temperature results of Fig. 9. Since the energy level spacings of the quartic oscillator increase with increasing energy, however, there will be a temperature dependence in the frequency of the dipole autocorrelation function. The nature of the temperature dependence will be to decrease the effective period of $G_{xx}(t)$ with increasing temperature. It is gratifying to see that the SPMC method accurately reproduces both the temperature dependence of the frequency and the dephasing observed at elevated temperatures.

F. Direct Time Correlation Functions

1. Formal Development

In the previous section we developed a number of methods to evaluate thermally symmetrized correlation functions. As indicated, the thermally symmetrized correlation functions contain the same thermodynamic information as the more familiar direct time correlation functions, but with a symmetric structure. This symmetric structure was seen to provide a combination of computational and formal advantages to the evaluation and analysis of dynamical properties of interacting quantum many-body systems. Of great importance was the ease at which classical limits become evident within the language of the symmetrized correlation functions. This classical limit was made clear in Section III.B.

In spite of these formal and computational advantages, symmetrized correlation functions can prove to be deficient in a number of circumstances. First, it is the direct time correlation functions that are related to physical properties directly. For example, in the case that the operators A and B in Eqs. (3.1) and (3.3) are equal and set to the dipole operators, the Fourier transform of the direct correlation function (Eq. 3.3) provides the infrared spectrum of a many-particle system within the dipole approximation. Of course, as indicated in Eq. (3.4), the Fourier transform of $C_{AA}(t)$ can be obtained from $G_{AA}(t)$. Unfortunately, as is evident from Eq. (3.4), at low temperatures $G_{AA}(t)$ becomes very small because of the exponentially decaying thermal factors. Indeed, we have found that $G_{AA}(t)$ cannot be computed from analytic expressions for an oscillator when the temperature becomes sufficiently low. Furthermore, if a calculation of $C_{AA}(\omega)$ were attempted from Eq. (3.4) at low temperatures using Monte Carlo methods, the thermal factors can be expected to magnify the errors, making the resulting computation nearly valueless. It is clear from the point of view of practical computation that it is important to find methods for the direct calculation of $C_{AA}(t)$. In this section methods are developed for the calculation of the direct time correlation functions. These methods parallel those developed for the symmetrized correlation functions. However, some distinct features will be evident for the direct time correlation functions that are not found in the symmetrized counterparts.

We begin the development with the defining equation expressed in Eq. (3.3). As with the symmetrized correlation functions, we express the quantum mechanical traces in coordinate representation. However, by virtue of the cyclic invariance of quantum mechanical traces, Eq. (3.3) can also be written

$$C_{AB}(t) = \frac{\text{tr}[e^{-\beta_c H} A e^{iHt/\hbar} B]}{\text{tr}[e^{-\beta H}]}, \tag{3.61}$$

where in this instance

$$\beta_c = \beta + it/\hbar. \tag{3.62}$$

Note that here β_c contains β rather than $\beta/2$ as in Eq. (3.2). Using the two expressions Eqs. (3.3) and (3.61), and evaluating the traces in coordinate representation, the two separate expressions for $C_{AB}(t)$ which follow can be readily derived, namely,

$$C_{AB}(t) = \frac{\int dx \, dx' \, dx'' \langle x| e^{-\beta H} |x'\rangle \langle x'| e^{iHt/\hbar} |x''\rangle \langle x''| e^{-iHt/\hbar} |x\rangle A(x')B(x'')}{\int dx \, dx' \, dx'' \langle x| e^{-\beta H} |x'\rangle \langle x'| e^{iHt/\hbar} |x''\rangle \langle x''| e^{-iHt/\hbar} |x\rangle} \tag{3.63}$$

and

$$C_{AB}(t) = \frac{\int dx \, dx' \langle x| e^{iHt/\hbar} |x'\rangle \langle x'| e^{-\beta_c H} |x\rangle A(x)B(x')}{\int dx \, dx' \langle x| e^{iHt/\hbar} |x'\rangle \langle x'| e^{-\beta_c H} |x\rangle}. \tag{3.64}$$

Of the two expressions we have found Eq. (3.64) to be preferable, and we have limited our numerical experience to its evaluation. Although Eq. (3.63) is more symmetric in structure than Eq. (3.64), the smaller number of integration variables in Eq. (3.64) currently seems more suitable to numerical study. However, our experience at this writing is preliminary, and further study of Eq. (3.63) is certainly appropriate. In the current discussion we limit consideration to Eq. (3.64).

As with the development of Eq. (3.24), after the introduction of Fourier path integral expressions for the matrix elements appearing in Eq. (3.64), the expression for the direct time correlation function can be shown to be

$$C_{AB}(t) = \frac{\int dx \, dx' \, d\mathbf{a} \, d\mathbf{b} \, A(x)B(x')\rho(x, x', [\mathbf{a}]) \exp\{i\tau\phi(x, x', [\mathbf{a}], [\mathbf{b}])\}}{\int dx \, dx' \, d\mathbf{a} \, d\mathbf{b} \, \rho(x, x', [\mathbf{a}]) \exp\{i\tau\phi(x, x', [\mathbf{a}], [\mathbf{b}])\}}, \tag{3.65}$$

where in this case the ratio of the real time to thermal time τ is given by

$$\tau = t/\beta\hbar. \tag{3.66}$$

In Eq. (3.65) the expression for the density function is given by

$$\rho(x, x', [\mathbf{a}]) = \exp\left\{ -\frac{m(x - x')^2\beta}{2\hbar^2|\beta_c|^2} - \sum_{k=1}^{\infty} \frac{a_k^2}{2s_k^2} - \beta\langle V\rangle_{\mathbf{a}} \right\}, \tag{3.67}$$

and the phase function is given by

$$\phi(x, x', [\mathbf{a}], [\mathbf{b}]) = -\frac{m(x - x')^2 \beta}{2\hbar^2 |\beta_c|^2 \tau^2} + \sum_{k=1}^{\infty} \frac{a_k^2}{2s_k^2} + \beta[\langle V \rangle_\mathbf{b} - \langle V \rangle_\mathbf{a}]$$

$$-\frac{|\beta_c|^2}{\beta^2 \tau^2} \sum_{k=1}^{\infty} \frac{b_k^2}{2s_k^2}. \tag{3.68}$$

In analogy with the development of the symmetrized time correlation functions we have put

$$s_k^2 = \frac{2\hbar^2 |\beta_c|^2}{m\beta(k\pi)^2} \tag{3.69}$$

The notation $\langle V \rangle_\mathbf{a}$ and $\langle V \rangle_\mathbf{b}$ is identical to the notation introduced for the symmetrized correlation functions. In contrast to symmetrized correlation functions, from Eq. (3.67) it is evident that the weight function ρ is a function of only one of the sets of Fourier coefficients.

In the case of symmetrized correlation functions it was seen that difficulties occurred at large values of τ owing to the large oscillations occurring in the phase function. From Eq. (3.68) it is clear that the same difficulties can be expected in the case of the direct time correlation functions. However, in contrast to the symmetrized case, for direct time correlation functions there is also difficulty at very short times because of the factors of τ appearing at the denominators of the first and last terms on the right-hand side of Eq. (3.68). Of course at $t = 0$ there is no difficulty because the calculation can be expressed as an equilibrium average. For both short (but finite) and long times, stationary phase Monte Carlo methods have proved to be as useful for direct time correlation functions as in the symmetrized case. To see how the stationary phase methods can be used, we introduce damping functions into the numerator and the denominator of Eq. (3.65) to obtain

$$C_{AB}(t)$$

$$= \frac{\int dx\, dx'\, d\mathbf{a}\, d\mathbf{b}\, A(x)B(x')\rho(x, x', [\mathbf{a}])D_\varepsilon(x, x', [\mathbf{a}], [\mathbf{b}]) \exp\{i\tau\phi(x, x', [\mathbf{a}], [\mathbf{b}])\}}{\int dx\, dx'\, d\mathbf{a}\, d\mathbf{b}\, \rho(x, x', [\mathbf{a}])D_\varepsilon(x, x', [\mathbf{a}], [\mathbf{b}]) \exp\{i\tau\phi(x, x', [\mathbf{a}], [\mathbf{b}])\}},$$

$$\tag{3.70}$$

where the damping function is given by

$$D_\varepsilon(x, x', [\mathbf{a}], [\mathbf{b}]) = \int d\mathbf{a}'\, d\mathbf{b}'\, \delta_\varepsilon([\mathbf{a}'], [\mathbf{b}']) \frac{\rho(x, x', [\mathbf{a} - \mathbf{a}'])}{\rho(x, x', [\mathbf{a}])},$$

$$\times \exp(i\tau\{\phi(x, x', [\mathbf{a} - \mathbf{a}'], [\mathbf{b} - \mathbf{b}']) - \phi(x, x', [\mathbf{a}], [\mathbf{b}])\}). \tag{3.71}$$

In Eq. (3.71) $\delta_\varepsilon([\mathbf{a}], [\mathbf{b}])$ is any prelimit form of a delta function. As in the case of the symmetrized correlation functions, we have found it convenient to choose $\delta_\varepsilon([\mathbf{a}], [\mathbf{b}])$ to be a multidimensional gaussian whose first moments are all zero and whose second moments are uncorrelated. Using the aforementioned choice for the weight function, we can then approximately express the phase in a first-order gradient expansion to obtain the zeroth order damping function

$$D_{0,\varepsilon}(x, x', [\mathbf{a}], [\mathbf{b}]) = \exp\left(-\frac{1}{2} \sum_{k=1}^{\infty} \tau^2 \varepsilon_k^2 \left\{ \left(\frac{\partial \phi}{\partial a_k}\right)^2 + \left(\frac{\partial \phi}{\partial b_k}\right)^2 \right\} \right). \qquad (3.72)$$

As in the symmetrized case, the direct time correlation function can be expressed as the ratio of two averages. We have found it best to evaluate the averages with respect to the density function $\rho(x, x', [\mathbf{a}]) D_{0,\varepsilon}(x, x', [\mathbf{a}], [\mathbf{b}])$ to obtain

$$C_{AB}(t) \approx \frac{\langle A(x)B(x') \exp\{i\tau\phi(x, x', [\mathbf{a}], [\mathbf{b}])\} \rangle}{\langle \exp\{i\tau\phi(x, x', [\mathbf{a}], [\mathbf{b}])\} \rangle}. \qquad (3.73)$$

2. Applications

In previous sections we have illustrated the methods developed with calculations on the quartic oscillator. This example can serve as a reference for those wishing to test the methods computationally. In this section we illustrate the application of the methods developed in Section III.F.1 on the harmonic oscillator rather than the quartic oscillator. We have chosen to use the harmonic rather than the quartic oscillator for two reasons. The first is that the existence of multiple stationary phase points at intermediate to long times for the quartic oscillator considerably complicates the numerical difficulties in applying the methods. Techniques for incorporating multiple stationary phase points in an optimal way are still being developed at this writing, and we feel it would be premature for us to emphasize any of the alternatives in this review. Multiple stationary phase points do not occur for the harmonic oscillator. The second reason for choosing the harmonic oscillator as an example system is that the analytical expressions we can derive for the oscillator illustrate some potential pitfalls in calculations on direct time correlation functions which are not evident in the pure numerical study which would be imposed by an investigation of the quartic oscillator.

We begin the formal treatment by giving the exact dipole autocorrelation function for a harmonic oscillator of mass m and natural frequency ω at temperature T,

$$C_{xx}(t) = \frac{\hbar}{2m\omega \sinh(\frac{1}{2}\beta\hbar\omega)} \left[\cos(\omega t)\cosh\left(\frac{1}{2}\beta\hbar\omega\right) + i\sin(\omega t)\sinh\left(\frac{1}{2}\beta\hbar\omega\right) \right].$$

$$(3.74)$$

It is immediately evident from Eq. (3.74) that in contrast to symmetrized correlation functions, the direct time correlation functions are complex rather than real. As is usual for calculations using the Fourier path integral method, the propagator matrix elements are truncated at a finite number of Fourier coefficients, k_{max}. If the averages in the numerator and denominator of Eq.(3.65) are evaluated with propagator matrix elements evaluated with a truncated number of Fourier coefficients, the direct time dipole autocorrelation function is given approximately by

$$C_{xx}^{k_{max}}(t) = \frac{1}{2}\frac{\zeta}{\zeta^2 - \kappa^2},$$

$$(3.75)$$

where

$$\kappa = \frac{m\beta}{2\hbar^2|\beta_c|^2} + i\frac{m\beta^2}{2\hbar t|\beta_c|^2} + \frac{\beta m\omega^2}{6} - \sum_{k=1}^{k_{max}}\left(\frac{E_k^2}{4F_k} + \frac{H_k^2}{4L_k}\right),$$

$$(3.76)$$

and

$$\zeta = -\frac{m\beta}{2\hbar^2|\beta_c|^2} - i\frac{m\beta^2}{2\hbar t|\beta_c|^2} + \frac{\beta m\omega^2}{12} + \sum_{k=1}^{k_{max}}(-1)^k\left[\frac{E_k^2}{4F_k} + \frac{H_k^2}{4L_k}\right]$$

$$(3.77)$$

In Eqs. (3.76) and (3.77) we have set

$$E_k = \frac{\beta m\omega^2}{k\pi} + i\frac{tm\omega^2}{\hbar k\pi},$$

$$(3.78)$$

$$F_k = \frac{1 + (|\beta_c|\hbar\omega/k\pi)^2}{2s_k^2} - \frac{it}{\beta\hbar}\frac{1 - (|\beta_c|\hbar\omega/k\pi)^2}{2s_k^2},$$

$$(3.79)$$

$$L_k = i\left[\frac{\hbar|\beta_c|^2}{2t\beta s_k^2} - \frac{tm\omega^2}{4\hbar}\right],$$

$$(3.80)$$

and

$$H_k = \frac{itm\omega^2}{\hbar k\pi}.$$

$$(3.81)$$

From this analytic expression for the direct time dipole autocorrelation function for the harmonic oscillator it is now possible to identify one difficulty which arises that was absent in symmetrized correlation functions. It is easy to show that the imaginary parts of both κ and ζ become infinite as the physical time approaches $n\pi/\omega$. From Eq. (3.75) it is also possible to show that these infinite imaginary parts cancel when combined to form the direct time dipole autocorrelation function. This divergence occurs only for finite k_{max} and has no overall effect on the final result. However, we have found that at times exactly equal to $n\pi/\omega$, the numerical evaluation of $C_{xx}(t)$ using path integral Monte Carlo methods is seriously unstable. We have also found that for physical times differing from $n\pi/\omega$ by as little as 0.1%, the numerical difficulties associated with the divergences are not evident. Whether this divergent behavior is a problem only for the harmonic oscillator or whether it is a problem for all systems at, for example, turning points, is not currently known.

In addition to the analytic discussion of calculations of direct time correlation functions for the harmonic oscillator, we also present a numerical study[87] using Monte Carlo methods. In calculations of correlation functions using stationary phase Monte Carlo methods, a primary concern is the choice of the ε parameter in the damping function. Although the results are formally independent of ε when the exact damping function is used, the damping parameter must be chosen with care in calculations at any level. If an approximation to the true damping function is used (e.g., Eq. 3.72), the results will depend upon ε. If the exact damping function is constructed by a separate Monte Carlo calculation as in Eq. (3.51), the choice of the damping parameter will govern the rate of convergence of the corrections to the approximate damping function with respect to the number of Monte Carlo points included. Consequently, we need to gain some intuition concerning the dependence of the results on the damping parameter. In Fig. 10 we plot the value of $\text{Re}[C_{xx}(\omega t/2\pi = 3.4)]$ as a function of ε for a harmonic oscillator. The temperature in the calculation was taken to be $\beta\hbar\omega = 2$ and the results were evaluated from Eq. (3.73) using Eq. (3.72) as the damping function. The mass of the oscillator was taken to be the mass of a hydrogen atom. The horizontal line is the exact result and the calculated points are depicted with associated Monte Carlo error bars when 10 million Monte Carlo points were included in the calculation. Eight Fourier coefficients were included in the evaluation of the propagator and density matrix elements. From the figure it is evident that the fluctuations in the Monte Carlo calculations are very large until a critical value of ε is reached. In the current calculation the critical value of the damping parameter is approximately $\varepsilon = 0.1$. Above the critical value of the damping parameter, the fluctuations in the calculations are roughly independent of ε. It is also evident that at the critical value of the damping parameter the systematic error resulting from the use of an approximate

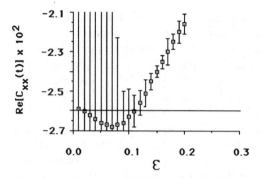

Figure 10.   The real part of $C_{xx}(t)$ (au) for the harmonic oscillator example of Section III at $\omega t/2\pi = 3.4$ as a function of the SPMC damping parameter ε. All calculations used a first-order gradient approximation for the damping function (Eq. 3.72), eight Fourier coefficients, and a common number of Monte Carlo quadrature points (10^7). The exact $k_{max} = 8$ value for the time correlation function for this time and temperature ($\beta\hbar\omega = 2$) is denoted by the horizontal line in the figure.

damping function is rather small. From the oscillator results, we reach the empirical observation that in a calculation it may be generally prudent to assess the dependence of the variance of the results on the damping parameter. The optimal choice of the damping parameter can be expected to occur when the variance stabilizes. Although we cannot prove this conjecture, we have found it to be a reliable procedure in other one-dimensional systems where the results can be compared to either exact results or the results of basis set expansions.

In Fig. 11 we show the calculated values for $\text{Re}[C_{xx}(t)]$ for times out to 3.4 periods of oscillation using the system parameters identical to those in Fig. 10. The solid line is the exact correlation function obtained from Eq. (3.70). The calculated points along with their single standard deviation error bars are shown, and the number of Fourier coefficients included are also indicated. The near linear increase in the number of Fourier coefficients required with time is the same as what we observed in Section III when we discussed the symmetrized time correlation functions for the oscillator. For times as short as $\omega t/2\pi = 0.1$, where the phase function is quite large, no special numerical difficulties were encountered.

As we discussed, in extending stationary phase Monte Carlo methods to other physical systems, problems associated with multiple and possibly disconnected stationary phase regions can be expected. At this writing methods to handle these multiple stationary phase regions are an object of ongoing research, and we leave a discussion of proposed methods to the research literature, rather than this review.

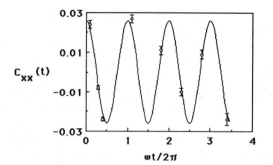

Figure 11. The real part of $C_{xx}(t)$ (au) for the harmonic oscillator example of Section III. The solid line is the exact result (Eq. 3.74). The results of SPMC calculations are shown. The number of Fourier coefficients used ranged from one at short times to eight at the largest time shown. These calculations were obtained using a first-order gradient approximation to the damping function (Eq. 3.72).

IV. SUMMARY AND DISCUSSION

The present review has summarized the application of Fourier path integral Monte Carlo methods to both equilibrium and dynamical problems in statistical mechanics. By presenting a variety of numerical examples, we have attempted to show the utility of the approach to both few- and many-body systems. That such a diverse range of problems and phenomenology can be expressed in a common language is an especially appealing feature of path integral approaches to many-body physics and chemistry. The value of this common language is illustrated by the recent work of Carraro and Koonin,[88] where the SPMC method has proved of use in problem areas quite distinct from those for which it was originally developed. In addition to providing a unified description of many-body phenomena, path integral methods provide a useful basis for the construction of practical and generalizable algorithms for numerical study. The importance of these techniques is likely to grow with advances in computer technology. In particular, path integral Monte Carlo methods are ideally suited to take advantage of parallel computer architectures.

Because path integral methods remain an area of current research, we wish to conclude by identifying a number of areas where work remains to be completed at this writing. In both equilibrium and dynamical applications the incorporation of particle statistics remains a problem. Useful algorithms have been developed for bosons.[16] Analogous algorithms for fermions are the subject of much current attention,[3,17] and a general solution is not yet available. Within the equilibrium domain, it would be useful to develop simple kinetic energy estimators that do not suffer the variance difficulties present in some of the current approaches.

As in classical Monte Carlo methods, a number of issues concerning sampling strategies are present in path integral approaches. The quantum mechanical problem is inherently a more difficult one than its classical counterpart. The auxiliary degrees of freedom introduced to describe the quantum mechanical paths introduce a multitude of length scales not present in the classical discussion. This problem is most severe in equilibrium applications at low temperatures, where the number of auxiliary degrees of freedom required for convergence will grow without limit as the temperature approaches zero. For any temperature, this difficulty is present in quantum dynamics at sufficiently long times. An additional concern in dynamical applications relates to possible sampling difficulties associated with the stationary phase importance sampling method. The tedious, if familiar, concern involves assuring that the stochastic walk underlying the Monte Carlo procedure adequately samples all relevant configuration space. It is appropriate to note that this problem is a generic Monte Carlo issue, not unique to the present application. As such, there exists a backlog of experience in related matters on which we can build. In stationary phase Monte Carlo applications we anticipate that our ability to characterize analytically the underlying stationary phase regions will prove of assistance. At this writing, these sampling issues are active fields of current research.

Acknowledgments

JDD wishes to thank Brown University and the Department of Chemistry for its generous support of this research. DLF acknowledges the donors of the Petroleum Research Fund of the American Chemical Society, grants from Research Corporation and the University of Rhode Island Academic Computer Center for partial support of this work. TLB wishes to acknowledge a grant of computer time from the Ohio Supercomputer Center as well as computational support from the University of Cincinnati Computer Center. We would like to thank Prof. R. D. Coalson for continuing discussions on the issues presented here. We would like to thank Drs. Don Frantz, David Leitner, and Steve Rick for helpful discussions concerning this chapter.

Appendix

As indicated in the text, the present methods readily generalize to many degrees of freedom. The difficulties involved are essentially notational in character. We define $x_{i,\alpha}$ as the αth component ($\alpha = x, y, z$) of the Cartesian position of particle i. In analogy with Eq. (2.34), paths that connect $x_{i,\alpha}$ and $x'_{i,\alpha}$ are written as

$$x_{i,\alpha}(u) = x_{i,\alpha} + (x'_{i,\alpha} - x_{i,\alpha})u + \sum_{k=1}^{\infty} a_{k,i,\alpha}\sin(k\pi u), \qquad (A.1)$$

where the dimensionless time variable u ranges from zero to unity. Rather than a single set of Fourier coefficients, we thus have an auxiliary set $(k = 1, k_{max})$ for each Cartesian component $(\alpha = x, y, z)$ of each particle $(i = 1, N)$ in the system. The necessary many-particle generalizations of the single variable results discussed in the text are easily obtained. For example, the general complex temperature density matrix element for an N-particle system with the interaction potential $V(\mathbf{x}_1, \mathbf{x}_2, \ldots, \mathbf{x}_N)$, is given by

$$\frac{\rho(\mathbf{x}_1, \mathbf{x}_2, \ldots, \mathbf{x}_N, \mathbf{x}_1', \mathbf{x}_2', \ldots, \mathbf{x}_N'; \beta_c)}{\rho_{fp}(\mathbf{x}_1, \mathbf{x}_2, \ldots, \mathbf{x}_N, \mathbf{x}_1', \mathbf{x}_2', \ldots, \mathbf{x}_N'; \beta_c)}$$
$$= \frac{\int d\mathbf{a}_1 \, d\mathbf{a}_2 \ldots d\mathbf{a}_N \exp(-\sum_{k,i,\alpha} a_{k,i,\alpha}^2 / 2\sigma_{k,i}^2 - \beta_c \langle V \rangle_a)}{\int d\mathbf{a}_1 \, d\mathbf{a}_2 \ldots d\mathbf{a}_N \exp(-\sum_{k,i,\alpha} a_{k,i,\alpha}^2 / 2\sigma_{k,i}^2)}, \qquad (A.2)$$

where

$$\sigma_{k,i}^2 = \frac{2\beta_c \hbar^2}{m_i \pi^2 k^2}, \qquad (A.3)$$

and where the temporal average of the potential along the path specified by Eq. (A.1) is given by

$$\langle V \rangle_a = \int_0^1 V(\mathbf{x}_1(u), \mathbf{x}_2(u), \ldots, \mathbf{x}_N(u)) \, du \qquad (A.4)$$

Using Eq. (A.2) for the complex temperature density matrix elements, we can readily obtain the many-particle generalizations of the expressions presented in Sections II and III. For example, the expression for the average potential energy within the canonical ensemble is given by

$$\langle V \rangle = \frac{\int d\mathbf{x}_1 \, d\mathbf{x}_2 \ldots d\mathbf{x}_N \, d\mathbf{a}_1 \, d\mathbf{a}_2 \ldots d\mathbf{a}_N \exp(-\sum_{k,i,\alpha} a_{k,i,\alpha}^2 / 2\sigma_{k,i}^2 - \beta_c \langle V \rangle_a) V(\mathbf{x}_1, \mathbf{x}_2, \ldots, \mathbf{x}_N)}{\int d\mathbf{x}_1 \, d\mathbf{x}_2 \ldots d\mathbf{x}_N \, d\mathbf{a}_1 \, d\mathbf{a}_2 \ldots d\mathbf{a}_N \exp(-\sum_{k,i,\alpha} a_{k,i,\alpha}^2 / 2\sigma_{k,i}^2)}.$$
$$(A.5)$$

References

1. See, for example, M. Kalos and P. A. Whitlock, *Monte Carlo Methods*, Wiley-Interscience, New York, 1986.
2. N. Metropolis, A. W. Rosenbluth, M. N. Rosenbluth, A. H. Teller, and E. Teller, *J. Chem. Phys.* **21**, 1087 (1953).
3. For a cross section of recent activity in this area, see the proceedings of the Metropolis quantum Monte Carlo conference, published as *J. Stat. Phys.* **43**, 729–1244 (1986).
4. Reference 1, pp. 157–168, and references therein. See also B. Alder and D. Ceperley, *Science* **231**, 555 (1986).

5. E. C. Behrman, G. A. Jongeward, and P. G. Wolynes, *J. Chem. Phys.* **79**, 6277 (1983).
6. V. S. Filinov, *Nucl. Phys.* **B271**, 717 (1986).
7. J. D. Doll, R. D. Coalson, and D. L. Freeman, *J. Chem. Phys*, **87**, 1641 (1987).
8. J. Chang and W. H. Miller, *J. Chem. Phys.* **87**, 1648 (1987).
9. J. D. Doll and D. L. Freeman, *Adv. Chem. Phys.* **73**, 289 (1989) and references therein.
10. J. D. Doll and L. E. Myers, *J. Chem. Phys.* **71**, 2880 (1979).
11. J. D. Doll and D. L. Freeman, *J. Chem. Phys.* **80**, 2239 (1984).
12. D. L. Freeman and J. D. Doll, *J. Chem. Phys.* **80**, 5709 (1984).
13. D. L. Freeman and J. D. Doll, *Adv. Chem. Phys.* **70 B**, 139 (1988) and references therein.
14. For a review of recent work, see B. J. Berne and D. Thirumalai, *Annu. Rev. Phys. Chem.* **37**, 401 (1986).
15. D. Thirumalai and B. J. Berne, *J. Chem. Phys.* **83**, 2972 (1985).
16. D. Ceperley and E. L. Pollack, *Phys. Rev. Lett.* **56**, 351 (1986).
17. For a cross section of recent activity, see, for example, *Quantum Monte Carlo Methods*, M. Suzuki (ed.), Springer-Verlag, Berlin, 1987.
18. For a recent review, see J. P. Valleau and S. G. Whittington in *Modern Theoretical Chemistry*, B. J. Berne (ed.), Plenum, New York, 1977, Vol. 5, pp. 137–168.
19. R. P. Feynman and A. R. Hibbs, *Quantum Mechanics and Path Integrals*, McGraw-Hill, New York, 1965.
20. For a discussion of various short-time propagators see N. Makri and W. H. Miller, *Chem. Phys. Lett.* **151**, 1 (1988).
21. R. P. Feynman, *Statistical Mechanics*, Benjamin, New York, 1972.
22. W. H. Miller, *J. Chem. Phys.* **63**, 1166 (1975).
23. J. A. Barker, *J. Chem. Phys.* **70**, 2914 (1979).
24. D. Chandler and P. G. Wolynes, *J. Chem. Phys.* **74**, 4078 (1981).
25. K. S. Schweizer, R. M. Stratt, D. Chandler, and P. G. Wolynes, *J. Chem. Phys.* **75**, 1347 (1981).
26. R. D. Coalson, *J. Chem. Phys.* **85**, 926 (1986).
27. J. D. Doll, R. D. Coalson, and D. L. Freeman, *Phys. Rev. Lett.* **55**, 1 (1985).
28. R. D. Coalson, D. L. Freeman, and J. D. Doll, *J. Chem. Phys.* **85**, 4567 (1986).
29. M. F. Herman, E. J. Bruskin, and B. J. Berne, *J. Chem. Phys.* **76**, 5150 (1982).
30. For a recent review of these methods and their applications see H. Risken, *The Fokker-Planck Equation; Methods of Solution and Applications*, Springer-Verlag, Berlin, 1984, Chapter 2.
31. N. Makri and W. H. Miller, *J. Chem. Phys.* **90**, 904 (1989).
32. R. D. Coalson, D. L. Freeman, and J. D. Doll, *J. Chem. Phys.* **91**, 4242 (1989).
33. E. P. Wigner, *Phys. Rev.* **40**, 749 (1932).
34. J. G. Kirkwood, *Phys. Rev.* **44**, 31, (1933).
35. Y. Fujiwara, T. A. Osborn, and S. F. J. Wilk, *Phys. Rev. A* **25**, 14 (1982).
36. D. L. Freeman and J. D. Doll, *J. Chem. Phys.* **82**, 462 (1985).
37. J. K. Lee, J. A. Barker, and F. F. Abraham, *J. Chem. Phys.* **68**, 1325 (1978).
38. T. L. Beck, J. D. Doll, and D. L. Freeman, *J. Chem. Phys.* **90**, 5651 (1989).
39. E. L. Pollack and D. Ceperley, *Phys. Rev.* **B30**, 2555 (1984).
40. D. Chandler, J. D. Weeks, and H. C. Andersen, *Science* **220**, 787 (1983) and references therein.
41. F. F. Abraham, *Homogeneous Nucleation Theory*, Academic, New York, 1974.
42. M. R. Hoare, P. Pal, and P. P. Wegener, *J. Colloid Interface Sci.* **75**, 126 (1980).
43. N. G. Garcia and J. M. S. Torroja, *Phys. Rev. Lett.*, **47**, 186 (1981).
44. R. D. Etters and J. B. Kaelberer, *Phys. Rev. A* **11**, 1068 (1975).
45. J. B. Kaelberer and R. D. Etters, *J. Chem. Phys.* **66**, 3223 (1977).
46. R. D. Etters and J. B. Kaelberer, *J. Chem. Phys.* **66**, 5112 (1977).
47. C. L. Briant and J. J. Burton, *J. Chem. Phys.* **63**, 2045 (1975).

48. D. J. McGinty, *J. Chem. Phys.* **58**, 4733 (1973).
49. W. D. Kristensen, E. J. Jensen, and R. M. J. Cotterill, *J. Chem. Phys.* **60**, 4161 (1974).
50. J. Jellinek, T. L. Beck, and R. S. Berry, *J. Chem. Phys.* **84**, 2783 (1986).
51. R. S. Berry, J. Jellinek, and G. Natanson, *Phys. Rev. A* **30**, 919 (1984).
52. H. Davis, J. Jellinek, and R. S. Berry, *J. Chem. Phys.* **86**, 6456 (1987).
53. J. D. Honeycutt and H. C. Andersen, *J. Phys. Chem.* **91**, 4950 (1987).
54. T. L. Beck, J. Jellinek, and R. S. Berry, *J. Chem. Phys.* **87**, 545 (1987).
55. T. L. Beck, J. D. Doll, and D. L. Freeman, *J. Chem. Phys.* **90**, 5651 (1989).
56. J. M. Ziman, *Principles of the Theory of Solids*, Cambridge University Press, Cambridge, UK, 1972, p. 63.
57. J. C. Shelley, R. J. LeRoy, and F. G. Amar, *Chem. Phys. Lett.* **159**, 14, (1988) and references therein.
58. J. Bosinger, R. Knochenmuss, and S. Leutwyler, *Phys. Rev. Lett.* **62**, 3058 (1989).
59. D. Eichenauer and R. J. LeRoy, *Phys. Rev. Lett.* **57**, 2920 (1986) and references therein.
60. M. Y. Hahn and R. L. Whetten, *Phys. Rev. Lett.* **61**, 1190 (1988).
61. See, for example, J. J. Erpenbeck and W. W. Wood in *Modern Theoretical Chemistry*, Vol. 6, B. J. Berne (Ed.), Plenum, New York, 1977, pp. 1–40; J. Kushick and B. J. Berne, *ibid.*, pp. 41–64.
62. M. Parrinello and A. Rahman, *J. Chem. Phys.* **80**, 860 (1984).
63. M. Sprik, M. Klein, and D. Chandler, *J. Chem. Phys.* **83**, 3042 (1985).
64. G. J. Maryna and B. J. Berne, *J. Chem. Phys.* **88**, 4516 (1988).
65. J. Schnitker and P. J. Rossky, *J. Chem. Phys.* **86**, 3471 (1987).
66. D. Scharf, J. Jortner, and U. Landman, *J. Chem. Phys.* **87**, 2716 (1987).
67. J. D. Doll, D. L. Freeman, and M. J. Gillan, *Chem. Phys. Lett.* **143**, 277 (1988).
68. J. D. Doll, T. L. Beck, and D. L. Freeman, *J. Chem. Phys.* **89**, 5753 (1988).
69. T. L. Beck, J. D. Doll, and D. L. Freeman, *J. Chem. Phys.* **90**, 3181 (1989).
70. N. Makri and W. H. Miller, *Chem. Phys. Lett.* **139**, 10 (1987).
71. N. Makri and W. H. Miller, *J. Chem. Phys.* **89**, 2170 (188).
72. N. Makri, *Chem. Phys. Lett.* **159**, 489 (1989).
73. B. J. Berne and C. D. Harp, *Adv. Chem. Phys.* **17**, 63 (1970).
74. W. H. Miller, S. D. Schwartz, and J. W. Tromp, *J. Chem. Phys.* **79**, 4889 (1983).
75. P. Schofield, *Phys. Rev. Lett.* **4**, 39 (1960).
76. P. A. Egelstaff and P. Schofield, *Nucl. Sci. Eng.* **12**, 260 (1962).
77. Reference 73, pp. 138–141.
78. See, for example, H. Goldstein, *Classical Mechanics*, Addison-Wesley, Reading, MA, 1981, p. 111.
79. W. H. Miller, *Adv. Chem. Phys.* **25**, 69 (1974), Eq. (6.3).
80. W. H. Miller, *Adv. Chem. Phys.* **30**, 77 (1975).
81. J. D. Doll, *J. Chem. Phys.* **81**, 3536 (1984).
82. See Appendix of Ref. 69.
83. L. R. Pratt, *J. Chem. Phys.* **85**, 5045 (1986).
84. J. D. Doll, T. L. Beck, and D. L. Freeman, *Int. J. Quantum Chem.*, **S23**, 73 (1989).
85. S. R Deans, *The Radon Transform and Some of its Applications*, Wiley-Interscience, New York, 1983.
86. See, for example, G. F. Carrier, M. Krook, and C. E. Pearson, *Functions of a Complex Variable*, McGraw-Hill, New York, 1966, pp. 272–275.
87. D. L. Freeman, T. L. Beck, and J. D. Doll in *Quantum Simulations of Condensed Matter Phenomena*, J. D. Doll and J. E. Gubernatis (eds.), World Scientific, Teaneck, NJ, 1990, pp. 58–70.
88. C. Carraro and S. E. Koonin, *Phys. Rev. B* **41**, 6741 (1990).

FLUCTUATING NONLINEAR HYDRODYNAMICS, DENSE FLUIDS, AND THE GLASS TRANSITION

BONGSOO KIM and GENE F. MAZENKO

The James Franck Institute and Department of Physics, The University of Chicago, Chicago, Illinois

CONTENTS

I. INTRODUCTION

In this chapter we discuss recent progress on the dynamics of dense fluids. We do not intend that this be a comprehensive review. Instead, we discuss the progress which resulted from the conjecture by Leutheusser[1] that certain mode-coupling nonlinearities lead to an ergodic–nonergodic transition[2] in very dense liquids which has many of the features of the liquid–glass transition. We will discuss this phenomena within the framework of fluctuating nonlinear hydrodynamics.[3-5] While we focus primarily on the recent work related to the glass transition, this work has roots in research of the 1960s and 1970s on nonequilibrium statistical mechanics. In the next section we will briefly review these developments which, in turn, puts the more recent work in perspective.

II. KINETIC THEORY AND MOLECULAR DYNAMICS

Following the lead of Maxwell,[6] Boltzmann,[7] and Chapman and Cowling,[8] the theory of the dynamics of simple fluids was dominated for almost 100 years by the kinetic theory of hard spheres. It is convenient to discuss the results of kinetic theory using the language[9,10] of equilibrium averaged time correlation functions:

$$C_{AB}(t) = \langle A(t)B(0)\rangle, \tag{2.1}$$

where A and B are dynamic variables of interest. Green[11]–Kubo[12] formula relate transport coefficients, λ, to integrals over current–current correlation functions

$$\lambda = \int_0^{+\infty} dt \langle J_\lambda(t)J_\lambda(0)\rangle, \tag{2.2}$$

where $J_\lambda(t)$ is the current associated with some conserved field. If, for example, J is the energy current, then λ is the thermal conductivity. If J is a component of the momentum current (stress tensor), then λ is one of the viscosities. A somewhat simpler example is the self-diffusion coefficient

$$D_s = \int_0^{+\infty} dt \, \psi(t), \tag{2.3}$$

where

$$\psi(t) = \langle \vec{V}_i(t) \cdot \vec{V}_i(0)\rangle / \langle \vec{V}_i^2(0)\rangle \tag{2.4}$$

is the normalized velocity autocorrelation function. In Eq. (2.4) $\vec{V}_i(t)$ is the velocity of the ith particle at time t.

For almost 100 years the only theory of transport phenomena in strongly interacting fluids was the Boltzmann equation (with Enskog corrections). The well-known basic physics of the Boltzmann equation includes short-range, uncorrelated, two-body collisions. Such a theory leads to correlation functions which decay exponentially with time:

$$\langle J_\lambda(t)J_\lambda(0)\rangle = \langle J_\lambda^2(0)\rangle e^{-t/\tau_\lambda}, \tag{2.5}$$

and associated transport coefficients, inserting Eq. (2.5) in Eq. (2.2), are given in terms of the appropriate decay rate:

$$\lambda = \langle J_\lambda^2(0)\rangle \tau_\lambda. \tag{2.6}$$

The development of the correlation function method in the 1960s allowed for a more systematic analysis of quantities like $\psi(t)$. Two approaches are notable: (1) the density expansion due to Zwanzig[13] and (2) the molecular dynamics method pioneered by Alder[14] and Wainwright. These two approaches complimented each other and led to some surprises. The density expansion for transport coefficients, invented by Zwanzig and implemented by a number of workers[15-18] showed that there are nonanalytic corrections to the Boltzmann equation result λ_B:

$$\lambda = \lambda_B(1 + A\rho + B\rho^2 \ln \rho + \cdots), \tag{2.6}$$

where ρ is the density and A and B are functions of temperature only. Initially the logarithmic term was interpreted as a divergence of the expansion. Eventually it was realized that the $\rho^2 \ln \rho$ term in Eq. (2.6) results from a "resummation" of so-called ring collision terms. Physically this meant that local collective behavior could serve to correlate (kick back) local few-particle dynamics and convert long time divergences in the Kubo formulas (Eq. 2.2) to finite nonanalytic density contributions. Molecular dynamics studies by Alder and Wainwright showed that correlation functions do *not* decay exponentially at long times. There are long-time tails of the form

$$\langle J_\lambda(t)J_\lambda(0)\rangle \sim A_\lambda t^{-d/2} \tag{2.7}$$

for large t and where d is the spatial dimensionality. These long-time tails can influence the associated transport coefficient in an important way. Let us define a time-dependent transport coefficient

$$\lambda(t) = \int_0^t d\tau \langle J_\lambda(\tau)J_\lambda(0)\rangle. \tag{2.8}$$

Clearly the physical transport coefficient is recovered as $t \to \infty$. Let us suppose the current correlation function can be approximated by Eq. (2.7) for times $t > t_1$. Then, for $t > t_1$, we can write Eq. (2.8) in the form

$$\lambda(t) = \lambda_s + A_\lambda \int_{t_1}^{t} d\tau\, \tau^{-d/2}$$

$$= \lambda_s + \frac{A_\lambda}{(1 - d/2)}(t^{1-d/2} - t_1^{1-d/2}), \qquad (2.9)$$

where λ_s is the contribution to λ over the time range 0 to t_1. For $d > 2$ we can take the long-time limit and the long-time tail makes the contribution $A_\lambda t_1^{(1-d/2)}/(d/2 - 1)$ to the transport coefficient. However, for $d \leq 2$ we have the spectacular result that conventional hydrodynamics does not exist![17] More precisely, transport coefficients are not well defined in two or less spatial dimensions:

$$\lim_{t \to \infty} \lambda_{d \leq 2}(t) \to \infty. \qquad (2.10)$$

The key point for our purposes here is that there is a common origin for both effects: collective or hydrodynamic effects on a semimicroscopic level. For dense fluids there is a coupling of single particle motion to collective (hydrodynamic) motion. In the case of the velocity autocorrelation function Alder and Wainwright[19] showed by direct inspection of their simulations that single particle motion could be enhanced by the particles being pushed by a semimacroscopic vortex motion. This persistent motion leads to the long-time tail in $\psi(t)$ and an enhancement of D_S. In the density expansion for transport coefficients one finds that the $\rho^2 \ln \rho$ contribution is from a collective screening in the multiple scattering process.

The results of these discoveries was the understanding that kinetic theory must be generalized to include the coupling of microscopic particle motion to collective hydrodynamic motions. One successful theory for amalgamating these two types of processes was developed by Mazenko[20] and implemented by Mazenko and Yip,[21] and Sjölander and Sjögren.[22] Related theories were proposed by Dorfman and Cohen,[23] and Resibois.[24] The results of these theories showed a tendency for the self-diffusion coefficient to decrease at high densities as shown in Fig. 1. It was tempting to think one was seeing a precursor to glass formation or crystalization where $D_s \to 0$. The theory was stuck at this point for some time. The reason for this impasse was that in the kinetic theory formalism, where one begins with the low density limit as a zeroth order approximation, it is difficult to extend the analysis into the dense

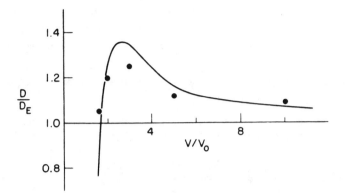

Figure 1. Self-diffusion coefficient ratio D/D_E versus density in a hard-sphere liquid, calculation by Furtado et al. (points) and molecular dynamics results (curve) (from ref. 21).

liquid regime. By the time one gets to the densities where D_s turns over in Fig. 1, one is losing control over the approximations made in the theory.

III. STOCHASTIC MODELS AND GENERALIZED HYDRODYNAMICS

Developed in parallel to the kinetic theory approach to nonequilibrium statistical mechanics was the method of fluctuating hydrodynamics. This approach evolved from very early work of Einstein,[25] Langevin,[26] Fokker[27] and Planck,[28] and Smoluchowski.[29] We will discuss the fluid version of this approach in considerable detail below. We begin, however, with a slightly more general point of view.

The general assumptions used in developing the theory of stochastic models and, more specifically, fluctuating hydrodynamics, is that there is a separation of time scales in the problem of interest and one restricts one's detailed attention to these variables which are evolving more slowly in time. Let us list the basic assumptions used in the approach.

1. One first selects the "slow modes" $\psi_i(\vec{x}, t)$ ($i = 1, 2, \ldots, m$) out of all the $6N$ microscopic degrees of freedom in a macroscopic (N-particle) system. The general criteria for selecting these degrees of freedom are to include:

   a. Conserved densities (mass density, momentum density, and energy density in a fluid).

   b. The Nambu–Goldstone transverse modes[30,10] associated with a broken (or nearly broken) continuous symmetry (transverse sound in a solid, spin waves in magnets, second sound in liquid crystals).

c. Modes that are slow because of a small parameter (motion of a large Brownian particle).[10]

2. Average over the $6N$-m fast variables to obtain the effective equation of motion satisfied by the slow variables.

Following the formal development of Zwanzig[31] and Mori[32] one can show, without approximation, that the Fourier transform of the slow modes

$$\psi_i(\vec{q}, t) = \int \frac{d^d x}{\sqrt{V}} e^{+i\vec{q} \cdot \vec{x}} \psi_i(\vec{x}, t), \tag{3.1}$$

where V is the volume of the system, satisfies the linear Langevin equation:

$$\frac{\partial \psi_i(\vec{q}, t)}{\partial t} = -i \sum_j \Omega_{ij}(\vec{q}) \psi_j(\vec{q}, t) - \sum_j \int_0^t d\bar{t}\, \Gamma_{ij}(\vec{q}, t - \bar{t}) \psi_j(\vec{q}, \bar{t}) + \eta_i(\vec{q}, t), \tag{3.2}$$

where $\Omega_{ij}(\vec{q})$ is called the collective frequency and is given in terms of static equilibrium correlation functions. We define the ψ_i such that $\langle \psi_i \rangle = 0$. The damping matrix, $\Gamma_{ij}(\vec{q}, t)$, consists of time-dependent transport coefficients and $\eta_i(\vec{q})$ is the **noise** which is related to $\Gamma_{ij}(\vec{q}, t)$ by

$$\langle \eta_i(\vec{q}, t) \eta_j(-\vec{q}, t') \rangle = 2k_B T \Gamma_{ij}(\vec{q}, t - t') \tag{3.3}$$

and satisfies

$$\langle \eta_i(\vec{q}, t) \psi_j(-\vec{q}, 0) \rangle = 0 \qquad t > 0. \tag{3.4}$$

One can obtain microscopic expressions for $\Omega_{ij}(\vec{q})$ and $\Gamma_{ij}(\vec{q}, t)$ using either the projection–operator method of Zwanzig and Mori or the resolvent operator method of Mazenko.[33]

If we multiply Eq. (3.2) by $\psi_j(-\vec{q}, 0)$ and average, we obtain, remembering Eq. (3.4),

$$\frac{\partial}{\partial t} C_{ij}(\vec{q}, t) = -i \sum_k \Omega_{ik}(\vec{q}) C_{kj}(\vec{q}, t) - \sum_k \int_0^t d\bar{t}\, \Gamma_{ik}(\vec{q}, t - \bar{t}) C_{kj}(\vec{q}, \bar{t}) = 0, \tag{3.5}$$

where

$$C_{ij}(\vec{q}, t) = \langle \psi_i(\vec{q}, t) \psi_j(-\vec{q}, 0) \rangle. \tag{3.6}$$

It is useful to introduce Laplace transforms

$$C_{ij}(\vec{q}, \omega) = -i \int_0^\infty dt\, e^{+i\omega t} C_{ij}(\vec{q}, t), \tag{3.7}$$

where $\text{Im}\,\omega > 0$, then Eq. (3.5) takes the convenient form

$$\sum_k [\omega \delta_{ik} - \Omega_{ik}(\vec{q}) - \Gamma_{ik}(\vec{q}, \omega)] C_{kj}(\vec{q}, \omega) = \chi_{ij}(\vec{q}), \tag{3.8}$$

where $\Gamma_{ik}(\vec{q}, \omega)$ is the Laplace transform of $\Gamma_{ik}(\vec{q}, t)$ and

$$\chi_{ij}(\vec{q}) = \langle \psi_i(\vec{q}) \psi_j(-\vec{q}) \rangle \tag{3.9}$$

is the static equal-time correlation function.

Conventional hydrodynamics results from the assumption that the small q and ω limit for the $\Gamma_{ik}(\vec{q}, \omega)$ is regular:

$$D_{ik} = \lim_{\vec{q}, \omega \to 0} \frac{1}{q^2} \Gamma_{ik}(\vec{q}, \omega). \tag{3.10}$$

There are a number of interesting circumstances where conventional hydrodynamics breaks down. We have already mentioned the case of long-time tails and the result that the limit (Eq. 3.10) does not exist in two or less dimensions. Other interesting examples are the breakdown of conventional hydrodynamics in isotropic antiferromagnets in the ordered phase[34] and in smectic A liquid crystals.[35] In three-dimensional fluids conventional hydrodynamics is well defined but because of the long-time tails there is a substantial contribution to the frequency-dependent viscosities.

Let us now specialize to the case of isotropic fluids and include in our set of slow variables the particle density $\rho(\vec{x}, t)$ and the momentum density $\vec{g}(\vec{x}, t)$. For simplicity we treat the isothermal case where we do not include the energy density. We do not expect that the inclusion of the energy will change any of the main features of our analysis. The static correlation functions in this case are easily found to be

$$\chi_{\rho\rho}(\vec{q}) = S(\vec{q}), \tag{3.11a}$$

where $S(\vec{q})$ is the static structure factor,

$$\chi_{\rho g_\alpha}(\vec{q}) = 0, \tag{3.11b}$$

and

$$\chi_{g_\alpha g_\beta}(\vec{q}) = \delta_{\alpha\beta} k_B T \rho_0, \tag{3.11c}$$

where ρ_o is the average mass density. Equation (3.11c) is essentially the equipartition theorem. In this case the collective frequency matrix is given by

$$\Omega_{\rho g_\alpha}(\vec{q}) = q_\alpha, \tag{3.12a}$$

$$\Omega_{g_\alpha \rho}(\vec{q}) = q_\alpha \rho_o k_B T S^{-1}(\vec{q}) \equiv q_\alpha c^2(\vec{q}), \tag{3.12b}$$

$$\Omega_{\rho\rho}(\vec{q}) = \Omega_{g_\alpha g_\beta}(\vec{q}) = 0, \tag{3.12c}$$

where α is a vector index and $c(\vec{q})$ is a generalization of the isothermal speed of sound to nonzero q. The matrix of transport coefficients is given by

$$\Gamma_{\rho\rho}(\vec{q}, \omega) = \Gamma_{\rho g_\alpha}(\vec{q}, \omega) = \Gamma_{g_\alpha \rho}(\vec{q}, \omega) = 0, \tag{3.13a}$$

$$\rho_0 \Gamma_{g_\alpha g_\beta}(\vec{q}, \omega) = \left[\frac{\eta(\vec{q}, \omega)}{3} + \zeta(\vec{q}, \omega) \right] q_\alpha q_\beta + \eta(\vec{q}, \omega) q^2 \delta_{\alpha\beta} \equiv \rho_0 q^2 D_{\alpha\beta}(\vec{q}, \omega), \tag{3.13b}$$

where $\eta(\vec{q}, \omega)$ is the generalized shear viscosity and $\zeta(\vec{q}, \omega)$ the generalized bulk viscosity. We can write $D_{\alpha\beta}(\vec{q}, \omega)$ as well as the current–current correlation function $C_{g_\alpha g_\beta}(\vec{q}, \omega)$ in terms of their longitudinal and transverse components:

$$D_{\alpha\beta}(\vec{q}, \omega) = \hat{q}_\alpha \hat{q}_\beta D_L(\vec{q}, \omega) + (\delta_{\alpha\beta} - \hat{q}_\alpha \hat{q}_\beta) D_T(\vec{q}, \omega), \tag{3.14}$$

$$C_{g_\alpha g_\beta}(\vec{q}, \omega) = \hat{q}_\alpha \hat{q}_\beta C_L(\vec{q}, \omega) + (\delta_{\alpha\beta} - \hat{q}_\alpha \hat{q}_\beta) C_T(\vec{q}, \omega), \tag{3.15}$$

where we note from Eq. (3.13b) that the generalized longitudinal viscosity is given by

$$D_L(\vec{q}, \omega) = \left[\frac{4}{3} \eta(\vec{q}, \omega) + \zeta(\vec{q}, \omega) \right] \Big/ \rho_0. \tag{3.16}$$

The Langevin equations become in this case

$$\frac{\partial \rho(\vec{q}, t)}{\partial t} = -i\vec{q} \cdot \vec{g}(\vec{q}, t), \tag{3.17a}$$

$$\frac{\partial g_\alpha(\vec{q}, t)}{\partial t} = -iq_\alpha c^2(\vec{q}) \rho(\vec{q}, t) - q^2 \sum_\beta \int_o^t D_{\alpha\beta}(\vec{q}, t - \bar{t}) g_\beta(\vec{q}, \bar{t}) d\bar{t} + \eta_\alpha(\vec{q}, t). \tag{3.17b}$$

Clearly Eq. (3.17a) is just the Fourier transform of the continuity equation and Eq. (3.17b) is the Fourier transform of the linearized fluctuating Navier–Stokes equation. Going over to the correlation function description given by Eq. (3.5) we can easily solve the coupled set of equations to obtain

$$C_{\rho\rho}(\vec{q}, \omega) = S(\vec{q}) \frac{\omega + iq^2 D_L(\vec{q}, \omega)}{\omega^2 - q^2 c^2(\vec{q}) + i\omega q^2 D_L(\vec{q}, \omega)}, \tag{3.18a}$$

$$C_{\rho g_L}(\vec{q}, \omega) = C_{g_L \rho}(\vec{q}, \omega) = \frac{k_B T}{m} \frac{q}{\omega^2 - q^2 c^2(\vec{q}) + i\omega q^2 D_L(\vec{q}, \omega)}, \tag{3.18b}$$

$$C_L(\vec{q}, \omega) = \frac{k_B T}{m} \frac{\omega}{\omega^2 - q^2 c^2(\vec{q}) + i\omega q^2 D_L(\vec{q}, \omega)}, \tag{3.18c}$$

and

$$C_T(\vec{q}, \omega) = \frac{k_B T}{m} \frac{1}{\omega + iq^2 \eta(\vec{q}, \omega)/\rho_o}. \tag{3.18d}$$

In the long wavelength low-frequency hydrodynamic limit $c(\vec{q})$ reduces to the isothermal speed of sound and we replace $\eta(\vec{q}, \omega)$ and $\zeta(\vec{q}, \omega)$ with the physical viscosities $\eta = \eta(\vec{o}, o)$ and $\zeta = \zeta(\vec{o}, o)$. In this limit, as shown in Fig. 2, we obtain the standard Brillouin sound doublet in the longitudinal correlation function and a diffusive peak in the transverse current correlation function.

This picture does not give any information about the local collective phenomena responsible for the long-time tails. Indeed this development is useful primarily to give the basic structure of the correlation functions for small q and ω and, using the microscopic expressions for the $D_{ij}(q, t)$ which

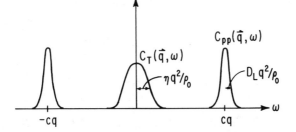

Figure 2.   The light-scattering spectrum of a simple liquid; Brillouin doublet and Rayleighs peak.

follow from the Zwanzig–Mori development, to derive the Green–Kubo relations

$$\frac{4}{3}\eta + \zeta = \int_0^{+\infty} dt \langle \sigma_{xx}(t)\sigma_{xx}(o) \rangle, \tag{3.19a}$$

and

$$\eta = \int_0^{+\infty} dt \langle \sigma_{xy}(t)\sigma_{xy}(o) \rangle, \tag{3.19b}$$

where $\sigma_{\alpha\beta}$ is $q \to 0$ limit of the Fourier transform of the microscopic stress tensor.

IV. FLUCTUATING NONLINEAR HYDRODYNAMICS

The stochastic equations of linearized hydrodynamics discussed in the last section are exact but beg the question of the local collective effects discussed in Section II. This physics is contained in the damping matrix $\Gamma_{ij}(\vec{q}, \omega)$. If there are long-time tails this will be reflected in the intermediate frequency behavior of $\Gamma_{ij}(\vec{q}, \omega)$. Within the formalism of linearized hydrodynamics $\Gamma_{ij}(\vec{q}, \omega)$ must be calculated using the microscopic correlation function expressions derived using the Zwanzig–Mori formalism.

A direct microscopic assault on the $\Gamma_{ij}(\vec{q}, \omega)$ can be made using kinetic theory methods as discussed in Section II. This is very difficult and it is not obvious how to develop systematic procedures for very dense systems. A less drastic approach is to develop an intermediate approach between kinetic theory and linear fluctuating hydrodynamics.

Let us pursue the following idea: If $\psi_i(\vec{x}, t)$ is a slow variable, should not $\psi_i^2(\vec{x}, t)$ also be a slow variable? The answer is yes and no. The reason the density, for example, is a slow variable is because it is conserved:

$$\lim_{q \to 0} \frac{\partial}{\partial t} \rho(\vec{q}, t) = 0.$$

This is *not* true for $\rho^2(\vec{x}, t)$ which satisfies

$$\lim_{q \to 0} \frac{\partial}{\partial t} \int d^3x \, e^{+i\vec{q}\cdot\vec{x}} \rho^2(\vec{x}, t) = 2 \int d^3x \, \rho(\vec{x}, t)(-\nabla \cdot \vec{g}(\vec{x}, t))$$

$$= 2 \int \frac{d^3k}{(2\pi)^3} i\vec{k} \cdot \vec{g}(\vec{k}, t)\rho(-\vec{k}, t).$$

Clearly this is not zero, but for long times the integral is dominated by the small k regime and the integral is small because of the factor of k. Thus it is sensible to look at a stochastic theory which includes $\psi_i(\vec{x}, t)$ *and* all products $\psi_i(\vec{x}, t)\psi_j(\vec{x}, t)$, $\psi_i(\vec{x}, t)\psi_j(\vec{x}, t)\psi_k(\vec{x}, t), \ldots$. The resulting formulation is known as *fluctuating nonlinear hydrodynamics*.

There are general pathways to the nonlinear equations governing fluids. The traditional approach, as discussed by Landau and Lifshitz[36] in their *Fluid Mechanics*, is to construct the equations of hydrodynamics using the conservation laws, constituitive relations, assumptions of local equilibrium, symmetry principles, and the restriction to long wavelengths. There is, however, a more general approach which also indicates more clearly the role of the thermal noise in the problem. This method[37] involves a generalization of the Langevin equation (3.2) to the case where $\psi_i(\vec{x}, t)$ is replaced by the variable

$$g_\phi(t) = \prod_R \prod_i \delta(\phi_i(\vec{R}) - \psi_i(\vec{R}, t)),$$

where $\psi_i(\vec{R}, t)$ is defined on a set of lattice points $\vec{R}$. As discussed by Ma and Mazenko[38] and Das and Mazenko,[4] one can derive a stochastic equation of motion for $g_\phi(t)$, which, in turn, generates a *nonlinear* Langevin equation for the ψ_i. We shall simply write down the general form of this equation here:

$$\frac{\partial \psi_\alpha(\vec{R}, t)}{\partial t} = V_\alpha[\vec{R}, t] - \int d^d R' \, \Gamma^0_{\alpha\beta}(\vec{R}, \vec{R}') \frac{\delta H}{\delta \psi_\beta(\vec{R}', t)} + \theta_\alpha(\vec{R}, t). \quad (4.1)$$

The first term on the right-hand side of Eq. (4.1) is the streaming velocity $V_\alpha[\vec{R}, t]$, which arises from the reversible part of the dynamics and is given explicitly by

$$V_\alpha[\vec{R}, t] = \int d^d R' \sum_\beta \{\psi_\alpha(\vec{R}, t), \psi_\beta(\vec{R}', t)\} \frac{\delta H}{\delta \psi_\beta(\vec{R}', t)}, \quad (4.2)$$

where $\{ \, , \, \}$ is the Poisson bracket. The quantity H appearing in the second term on the right-hand side of Eq. (4.1) and in Eq. (4.2) is the effective Hamiltonian appearing in the equilibrium probability distribution ($\sim e^{-\beta H}$) governing the slow variables ψ. The thermal noise θ_α in Eq. (4.1) is related to the "bare" set of dissipative coefficients $\Gamma^0_{\alpha\beta}(\vec{R}, \vec{R}')$ by

$$\langle \theta_\alpha(\vec{R}, t)\theta_\beta(\vec{R}', t') \rangle = 2k_B T \Gamma^0_{\alpha\beta}(\vec{R}, \vec{R}')\delta(t - t'). \quad (4.3)$$

In the case of isothermal fluids we have the set of slow variables $\psi = (\rho, \vec{g})$.

In order to generate the correct equations for isothermal fluids we must choose an effective Hamiltonian of the form[39]

$$H[\rho, \vec{g}] = H_K[\rho, \vec{g}] + H_u[\rho], \qquad (4.4)$$

where H_K is the kinetic energy term

$$H_K = \int d^3x \frac{1}{2}\vec{g}^2(\vec{x}, t)/\rho(\vec{x}, t) \qquad (4.5)$$

and H_u is a potential energy term dependent only on the density. The Poisson brackets among these variables are given by[4]

$$\{\rho(\vec{x}), \rho(\vec{x}')\} = 0, \qquad (4.6a)$$

$$\{\rho(\vec{x}), g_\alpha(\vec{x}')\} = -\nabla_x^\alpha[\delta(\vec{x} - \vec{x}')\rho(\vec{x})], \qquad (4.6b)$$

$$\{g_\alpha(\vec{x}), g_\beta(\vec{x}')\} = -\nabla_x^\alpha[\delta(\vec{x} - \vec{x}')g_\alpha(\vec{x})] + \nabla_{x'}^\beta[\delta(\vec{x} - \vec{x}')g_\beta(\vec{x})], \quad (4.6c)$$

and the Γ_{ij}^0 are identical in structure to those Γ_{ij} given by Eq. (3.13), except that the viscosities are replaced by their *bare* or local values η_0 and ζ_0. With these assumptions we obtain the standard nonlinear equations of hydrodynamics generalized to include a thermal noise term:

$$\frac{\partial \rho}{\partial t} = -\nabla \cdot \vec{g}, \qquad (4.7a)$$

$$\frac{\partial g_\alpha}{\partial t} = -\nabla_\alpha P[\rho] - \sum_\beta \nabla_\beta(\rho V_\alpha V_\beta) - \sum_\beta \hat{\Gamma}_{\alpha\beta}^0 V_\beta + \theta_\alpha. \qquad (4.7b)$$

The first equation is simply the equation of continuity. The second equation is the Navier–Stokes equation with the well-known terms. $-\nabla_\alpha P[\rho]$ is the reversible force term due to the gradient of the pressure. In this case the pressure is a nonlinear functional of the density which is related to the potential energy density

$$H_u = \int d^3x f[\rho(\vec{x})] \qquad (4.8)$$

by the usual thermodynamic relation

$$P = \rho \frac{\partial f}{\partial \rho} - f, \qquad (4.9)$$

where f has the physical interpretation as the free-energy per unit volume. The second term on the right of Eq. (4.7b) is the usual convective term where we have introduced the velocity field $\vec{V}(\vec{x}, t)$ defined by

$$\vec{g} = \rho \vec{V}. \tag{4.10}$$

The third term on the right is the viscous damping term where

$$\hat{\Gamma}^o_{\alpha\beta}(\vec{x}, \vec{x}') = -\left[\left(\frac{\eta_o}{3} + \zeta_o \right) \nabla_\alpha \nabla_\beta + \eta_o \delta_{\alpha\beta} \nabla^2 \right] \delta(\vec{x} - \vec{x}') \equiv \hat{\Gamma}^o_{\alpha\beta}(\vec{x}) \delta(\vec{x} - \vec{x}') \tag{4.11}$$

and η_o and ζ_o are now the bare shear and bulk viscosities. Finally θ_α is the thermal noise, assumed to be Gaussian, which satisfies

$$\langle \theta_\alpha(\vec{x}, t) \theta_\beta(\vec{x}', t') \rangle = 2 k_B T \hat{\Gamma}^o_{\alpha\beta}(\vec{x} - \vec{x}') \delta(t - t'). \tag{4.12}$$

It is easy to analyze Eq. (4.7) in the linear regime where

$$P[\rho] = c_o^2 \delta\rho, \tag{4.10}$$

$\delta\rho = \rho - \rho_o$, we can drop the convective terms, and

$$V_\beta = g_\beta / \rho_0. \tag{4.11}$$

In this limit Eq. (4.7) reduces in form to Eq. (3.17). The only difference is that the physical transport coefficient and speed of sound in the "exact" linear theory are replaced by the bare quantities η_o, ζ_o, and c_o^2. In the nonlinear theory we must resolve the full set of equations to obtain the physical transport coefficients and the speed of sound.

V. INCOMPRESSIBLE CASE

Our primary concern in the rest of this chapter will be to extract the physics contained in the set of equations given by Eq. (4.7). As we explain as we proceed this is not a simple task. Before discussing a direct assault on this set of equations, we discuss two partial solutions of interest. Let us first consider these equations in the incompressible limit where the density is taken to be a constant

$$\rho = \rho_o \tag{5.1}$$

and

$$\vec{g} = \rho_o \vec{V}. \tag{5.2}$$

From the continuity equation (Eq. 4.4a) we obtain

$$\nabla \cdot \vec{V} = 0 \tag{5.3}$$

and the velocity field is transverse. The Navier–Stokes equation (Eq. 4.4b) reduces in this case to

$$\rho_o \frac{\partial V_T^\alpha}{\partial t} = -\sum_{\beta, \gamma} P_T^{\alpha\gamma} \rho_o V_T^\beta \nabla_\beta V_T^\gamma + \eta_o \nabla^2 V_T^\alpha + \theta_T^\alpha \tag{5.4}$$

and $P_T^{\alpha\gamma}(\vec{x})$ is a transverse projection operator with Fourier transform

$$P_T^{\alpha\gamma}(\hat{k}) = \delta_{\alpha\gamma} - \hat{k}_\alpha \hat{k}_\gamma. \tag{5.5}$$

Equation (5.4) has been widely studied using a variety of sophisticated theoretical techniques.[40] For our purposes here the important result is that time-dependent physical viscosity obtained from Eq. (5.4) does show long-time tails

$$\eta(t) \sim A_\eta t^{-d/2}. \tag{5.6}$$

We note that this semimacroscopic calculation leads to the same exponent as that from a microscopic kinetic theory or molecular dynamics simulation. Next we observe that the coefficient A_η obtained in the analysis of Eq. (5.4) is "universal" in that it can be expressed in terms of physical quantities

$$A_\eta = \frac{(k_B T)}{15} \left(\frac{1}{8\pi} \right)^{3/2} \left(\frac{7}{\nu^{3/2}} + \frac{1}{\Gamma^{3/2}} \right). \tag{5.7}$$

What is surprising is that precisely the same coefficient was obtained by Resibois and Pomeau[17] in a very careful and definitive treatment of the fully microscopic kinetic theory. Thus the convective nonlinearity in the equation of fluctuating nonlinear hydrodynamics fully accounts for the long-time tails in the problem.

The convective nonlinearities in Eq. (5.4) lead to the breakdown of hydrodynamics in two or less spatial dimensions. As mentioned briefly above, there are other examples where nonlinear couplings in fluctuating nonlinear hydrodynamics leads to the breakdown of conventional hydrodynamics. In the ordered phase for isotropic antiferromagnets the spin diffusion coefficient diverges as q^{-1} as $q \to 0$ due to a nonlinear coupling between the longitudinal magnetization and the transverse components of the staggered magnetization. Similarly, three components of the viscosity tensor for smectic A liquid crystals diverge as ω^{-1} as $\omega \to 0$ due to a coupling between the momentum density and the layer displacement field. Similar nonlinear Langevin equations form the basis for our understanding of dynamic critical phenomena.[38,41]

VI. THE LEUTHEUSSER TRANSITION

Our general model, given by Eq. (4.4), can also be treated in another limit. Suppose we ignore the convective term in Eq. (4.4b) and assume $V_\beta = g_\beta/\rho_o$ in the damping term. Then our equations of motion take the form

$$\frac{\partial \rho}{\partial t} = -\nabla \cdot \vec{g}, \tag{6.1a}$$

$$\frac{\partial g_\alpha}{\partial t} = -\nabla_\alpha P[\rho] - \sum_\beta \hat{\Gamma}_{\alpha\beta} g_\beta/\rho_o + \theta_\alpha, \tag{6.1b}$$

and the only nonlinearities are in the pressure term. In terms of the general derivation of the equations (Eq. 4.4) the approximation (Eq. 6.1) corresponds to: (1) setting $\rho = \rho_o$ in the kinetic energy (Eq. 4.2) and replacing the Poisson bracket relation (Eq. 4.3) with the averages on the right-hand side. In this case the pressure is equal to the potential energy density. Let us assume

$$P[\rho] = c_o^2 \delta\rho + \frac{1}{2}\chi_o^{-1}(\delta\rho)^2 + \frac{1}{3}\lambda_3(\delta\rho)^3 + \cdots, \tag{6.2}$$

where $\rho = \rho_o + \delta\rho$. This is still a nontrivial model and we must be satisfied with developing a perturbation theory expansion in terms of the nonlinear terms. Generally speaking, such expansions are in powers of $(k_B T)^{1/2}$ since each factor of $\delta\rho$ or $\vec{g}$ can be associated with at least one factor of the noise which is proportional to $(k_B T)^{1/2}$. There are at least three ways of developing the perturbation theory for this set of equations. The most direct method, which involves the direct iteration of the Langevin equation, was developed by Ma and Mazenko[38] within the context of dynamic critical phenomena. This method is somewhat cumbersome with respect to the renormalization of the sound speed and the viscosity. The second approach uses a generalized Fokker–Planck method as developed in Mazenko, Nolan, and Freedman[34] and Mazenko, Ramaswamy, and Toner.[35] This method has the advantage of giving the equilibrium correlation functions directly and general expressions for the generalized transport coefficients. The disadvantage of the method is that it is difficult to work at higher than first order in the perturbation theory expansion. The third method, which we discuss in some detail in Section VIII, is the functional integral method originally due to Martin et al.[42] This method, like the generalized Fokker–Planck method, requires a great deal of formal development and it is difficult to extract the static equilibrium quantities directly. However, this method is the most powerful in developing renormalized perturbation theory at arbitrary order in the nonlinearities.

A direct perturbation theory expansion in powers of $(k_B T)^{1/2}$ can be constructed using any one of the three methods and each, as expected, gives the same result at lowest order. There is a renormalization of the sound speed which need not concern us here and a renormalized longitudinal viscosity,

$$D_L(\vec{q}, \omega) = D_0 + \frac{\chi_0^{-2}}{k_B T} \int_0^{+\infty} dt \, e^{+i\omega t} \int \frac{d^3k}{(2\pi)^3} C_{\rho\rho}(\vec{k}, t) C_{\rho\rho}(\vec{q} - \vec{k}, t). \quad (6.3)$$

Similar "mode-coupling" expressions have been derived directly from kinetic theory with the only modification that the product of $C_{\rho\rho}$'s in the integrand is multiplied by an "interaction vertex" $|V(\vec{q}, \vec{k})|^2$ (Mazenko,[20] Mazenko and Yip,[21] Leutheusser[43]). We now have a coupled nonlinear problem for $D_L(\vec{q}, \omega)$ together with the equation (3.18a) relating $C_{\rho\rho}(\vec{q}, \omega)$ and $D_L(\vec{q}, \omega)$.

Leutheusser[1] came to this set of equations from a different path. It is useful to paraphrase the physical picture behind his analysis. As the density increases in a dense fluid, the structural arrangement of the system becomes increasingly sluggish and hence the viscosities grow rapidly. Thus one has the development of a new very long time scale associated with the incipient freezing. This freezing on the local level will eventually propagate to arbitrarily long length scales and shift the hydrodynamic regime to much longer times and distances. Leutheusser assumed that this freezing or slowing down is driven by density fluctuations. With this motivation Leutheusser analyzed Eqs. (3.18a) and (6.3) using two simplifying assumptions.

1. Assume that $D_L(\vec{q}, \omega)$ is large, where the "general" hydrodynamical representation (Eq. 3.18a) becomes

$$\phi(\vec{q}, \omega) \equiv \frac{C_{\rho\rho}(\vec{q}, \omega)}{S(\vec{q})} = \frac{1}{\omega - [c^2(\vec{q})/iD_L(\vec{q}, \omega)]}. \quad (6.4)$$

2. Assume that $\phi(\vec{q}, \omega)$ and $D_L(\vec{q}, \omega)/c^2(\vec{q})$ are sufficiently weak functions of wavenumber that we can take them to be independent of wavenumbers and obtain the equations

$$\phi(\omega) = \frac{1}{\omega + i/d(\omega)}, \tag{6.5}$$

with $d(\omega) = D_L(\vec{q}, \omega)/c^2(\vec{q})$, and

$$d(\omega) = d_o + \lambda \int_0^{+\infty} dt \, e^{+i\omega t} \phi^2(t), \tag{6.6}$$

where

$$\lambda = k_B T \chi_0^{-2} \int \frac{d^3 k}{(2\pi)^3} S^2(\vec{k}) \tag{6.7}$$

and a large wavenumber cutoff is assumed. Equations (6.5) and (6.6) form Leutheusser's model.

The main feature of the solution to Eqs. (6.5) and (6.6) is the existence of an ergodic–nonergodic transition at a definite value λ_c of the coupling parameter λ. Thus in the liquid regime ($\lambda < \lambda_c$)

$$\lim_{t \to \infty} \phi(t) = 0, \tag{6.8}$$

while in the glass regime ($\lambda \geq \lambda_c$)

$$\lim_{t \to \infty} \phi(t) = h > 0. \tag{6.9}$$

The structure of the equations which allows for this transition can be seen analytically. If we simply set $\phi(t) = h$ in the Laplace transformed quantities, we obtain

$$\phi(\omega) = -i \int_0^{+\infty} dt \, e^{+i\omega t} h = \frac{h}{\omega}, \tag{6.10}$$

and in Eq. (6.6) we obtain

$$d(\omega) = d_0 + \frac{\lambda h^2}{-i\omega}. \tag{6.11}$$

Inserting Eqs. (6.10) and (6.11) in Eq. (6.5) we obtain

$$\frac{h}{\omega} = \frac{1}{\omega} \frac{\lambda h^2}{1 + \lambda h^2},$$ (6.12)

and the nonergodic solution is possible only if

$$1 = \frac{\lambda h}{1 + \lambda h^2}.$$ (6.13)

If we restrict our analysis to the transition point ($\lambda = \lambda_c$), we find the density correlation function has the form

$$\phi(t) = h_c + \phi_v(t),$$ (6.14)

$$\phi_v(t) = \phi_o t^{-\gamma} + \phi_1 t^{-2\gamma} + \cdots,$$ (6.15)

where λ_c, h_c, and γ are determined by matching the three divergent terms in d(ω),

$$d(\omega) = \frac{d_1}{\omega} + \frac{d_2}{\omega^{1-\gamma}} + \frac{d_3}{\omega^{1-2\gamma}},$$

where we assume $\gamma < 1/2$ and obtain Eq. (6.13) with $\lambda = \lambda_c$ and $h = h_c$ and the additional equations

$$1 + \lambda_c h_c^2 = (1 - h_c) 2\lambda_c h_c,$$ (6.16)

$$2h_c \Gamma^2 (1 - \gamma) = (1 - h_c) \Gamma(1 - 2\gamma),$$ (6.17)

where Γ is the gamma function. Equations (6.13) and (6.16) have the solution $\lambda_c = 4$ and $h_c = 1/2$ and γ is determined by the equation

$$\Gamma(1 - 2\gamma) = 2\Gamma^2(1 - \gamma).$$ (6.18)

The numerical solution of this equation gives $\gamma = 0.395$. Thus the density–density correlation function shows a power-law behavior in time for $\lambda = \lambda_c$. It is worth noting that the values for λ_c, h_c, and γ are not universal but depend on working at lowest nontrivial order in the expansion for D_L. The general structure of the solution to Eqs. (6.5) and (6.6) is richer than one might guess. The density correlation function has a general form

$$\phi(\omega) = \frac{h}{\omega + i\delta} + \phi_v(\omega),$$ (6.19)

where

$$
\delta = \begin{matrix} 0 & \lambda > \lambda_c \\ \delta_0 \varepsilon^{+\mu} & \lambda < \lambda_c \end{matrix} \tag{6.20}
$$

and $\varepsilon = |\lambda - \lambda_c| \sim |T - T_G| \sim |\rho - \rho_G|$. The exponent μ is not independent of γ but is given by

$$
\mu = (1 + \gamma)/2\gamma \approx 1.8. \tag{6.21}
$$

Analysis of the viscosities in the liquid $(\lambda < \lambda_c)$ phase shows a power-law divergence

$$
\eta \sim \Gamma_L \sim \varepsilon^{-\mu}. \tag{6.22}
$$

In the glass phase $(\lambda > \lambda_c)$ we obtain the results

$$
d(\omega) = i\frac{\lambda h^2}{\omega} + d_v(\omega), \tag{6.23}
$$

$$
\eta(\omega) = i\frac{\lambda_T h^2}{\omega} + \eta_v(\omega), \tag{6.24}
$$

where λ_T is different from λ, and

$$
d_v(0) \sim \eta_v(0) \sim \varepsilon^{-(\mu-1)}.
$$

Inserting the result (Eq. 6.24) for the shear viscosity into Eq. (3.18d) for the transverse current correlation function we obtain

$$
C_T(\vec{q}, \omega) = \frac{\rho_0 k_B T \omega}{\omega^2 - c_T^2 q^2 + iq^2 \omega \eta_v(\omega)}, \tag{6.25}
$$

where

$$
c_T^2 = 4\lambda_T h^2 \tag{6.26}
$$

and one has transverse sound in the "glass" phase, as one expects physically.

It is worth pointing out that the analytic analysis given above can be substantiated by a direct numerical analysis of Eqs. (6.5) and (6.6) in the time domain where

$$\left(1 + d_0 \frac{\partial}{\partial t}\right)\phi(t) + \lambda \int_0^t d\bar{t}\,\phi^2(t - \bar{t})\frac{\partial\phi(\bar{t})}{\partial\bar{t}} = 0. \tag{6.27}$$

We see in this case that the Leutheusser model maps onto a nonlinear oscillator with a retarded interaction.

VII. CRITIQUE OF THE LEUTHEUSSER TRANSITION

While the introduction of the Leutheusser mechanism raised expectations that an understanding of the glass transition was at hand, it also raised a number of important questions. A major practical question concerned the power-law temperature dependence of the viscosity predicted (Eq. 6.22) by the theory. It had been expected that viscosities in the vicinity of the glass transition should obey a Vogel[44]–Fulcher[45]

$$\eta = \eta_0 e^{A/(T - T_o)} \tag{7.1}$$

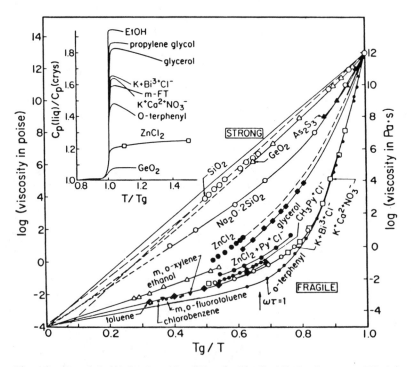

Figure 3. T_g-scaled plots for viscosities of glass-forming liquids of various types. T_g is defined as the temperature at which the viscosity reaches 10^{13} poise (from ref. 54).

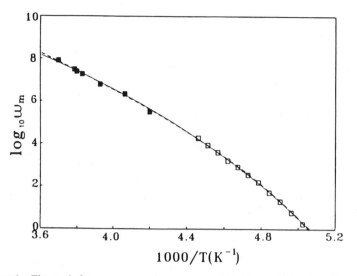

Figure 4. The peak frequency, on a log scale, vs. $1/T$. The solid line is a fit with the Vogel–Fulcher law: $\omega_m = 15.7 \times 10^{14} \exp[-2310/(T - 129)]$. The dotted line is a scaling law fit: $\omega_m = 15.7 \times 10^{10}[(T/175.5) - 1]^{12.5}$ (from ref. 47).

or Arrhenius

$$\eta = \eta_o e^{A/T} \tag{7.2}$$

form. Fits to such forms were known to describe data over nearly 20 decades of viscosity, as shown in Fig. 3.[46] Surprisingly it was found that power-law form Eq. (6.22) also gave good fits to the same data. An important point, however, was that the fitted exponents were very large[47] compared to the predictions of the theory (see Fig. 4). It is also important to note that analysis of experimental data[48] for the viscosity for a variety of dense fluids for temperatures well above the "glass transition" did show a power-law behavior with an exponent close to 2. This value is close to that found in the last section. Similarly, molecular dynamics experiments on several different types of systems showed a similar power-law behavior with an index near 2.[49]

A second related, but more philosophical, question was raised[50] concerning the possibility of a sharp dynamical transition occurring for a system with only a finite number of participating degrees of freedom. The Leutheusser transition is dynamical in that the sharp change in the viscosity as $T \to T_G$ is not accompanied by any changes in the thermodynamic properties of the system. The limited number of degrees of freedom are associated with the zero-spatial dimensional character of the Leutheusser equations, and the physical assumption that the glass transition is associated with spatially local behavior in the system. Thus one might intuitively argue that one can not rigorously have an ergodic–nonergodic transition at any nonzero temperature. While the system may become very sluggish, at any nonzero temperature without quenched disorder, the system should eventually equilibrate.

In addressing this question it is important to distinguish several separate points. The first point is that the Leutheusser equations, which depend only on a time variable, *do* show the dynamical transition. The transition is not an artifact of some approximate solution of these equations. Next, one must focus on whether the Leutheusser equation can be systematically derived from the microscopic equations governing a fluid. This is a delicate and difficult matter. If we start with the nonlinear equations (Eq. 6.1), it is easy to show, using generalized Fokker–Planck methods, that the pressure $P[\sigma]$ is generated from an effective potential energy density

$$P[\rho] = \rho_0 \frac{\delta}{\delta\rho} f, \qquad (7.3)$$

and nonlinearities in P result from higher than quadratic terms in H. We indicated in Section VI that a cubic term in f with a positive coefficient will generate the Leutheusser mechanism at first order in perturbation theory. However, there are classes of effective potential energy densities, such as

$$f(\rho) = \mu\delta\rho + \frac{r}{2}(\delta\rho)^2 + \frac{\lambda}{3}(\delta\rho)^3 + \frac{u}{4}(\delta\rho)^4 + \frac{1}{2}(\nabla\delta\rho)^2, \qquad (7.4)$$

where nonlinear kink solutions of the mean field equations

$$\frac{\delta f}{\delta\rho} = 0 \qquad (7.5)$$

give a mechanism for rapidly flipping the system from one local solution ρ_1^* to another ρ_2^*. Such disordering agents are missed by perturbation theory and should guarantee ergodicity in the associated dynamical model. Thus the Leutheusser transition would be destroyed by nonperturbative kink-like motions.

This argument (similar to the Landau argument against phase transitions in one-dimensional systems) is persuasive for those systems that support such static kink-like solutions, but is not definitive in the situation of interest here. One could try to think of systems which have single well potentials for all temperatures and which also generate the Leutheusser mechanism. This however seems to us to beg the question of whether (and which) fluids show the Leutheusser mechanism. At this point we should remember that Eq. (6.1) is an approximate form of the more general fluid equations (Eq. 4.4). We should recall the approximations in going from Eq. (4.4) and Eq. (6.1). The key points are a change in the basic Poisson brackets satisfied by ρ and $\vec{g}$, and this in turn leads to a quite different relation between the pressure and the potential energy functional. In the more general case of Eq. (4.4) the pressure is related to the effective potential energy density f via the standard thermodynamic relation (Eq. 4.6). We notice in this case a purely quadratic free-energy

$$f = \frac{1}{2}\chi_0^{-1}(\delta\rho)^2 \tag{7.6}$$

generates a nonlinear pressure

$$P = \rho\chi_0^{-1}\delta\rho - \frac{1}{2}\chi_0^{-1}(\delta\rho)^2,$$

$$= \rho_0\chi_0^{-1}\delta\rho + \frac{1}{2}\chi_0^{-1}(\delta\rho)^2. \tag{7.7}$$

Therefore, the equilibrium properties of this model should be completely benign! We do not expect any kink-like excitations for this system. The question remains: Does a realistic fluid show the Leutheusser transition? Does one find this transition in a careful analysis of the general equations (Eq. 4.6)?

It might appear that a serious shortcoming of our technique is the over-simplified treatment of the wavenumber dependence at intermediate wave-lengths. Our analysis in Section VI assumed that all wavelengths were equally important. Physically we expect that wavenumbers near the static structure factor maximum should be most important in understanding the glass transition.[51] The kinetic theory approach to the Leutheusser mechanism is able to treat the wavenumber dependence carefully and finds the same qualitative results found in Section VI. Similarly, Das[5] has generalized the stochastic model (Eq. 6.4) to include spatial structure with a peaked structure factor and again obtains the same qualitative result.

VIII. DAS AND MAZENKO ANALYSIS

A. The MSR Method

We now turn our attention to the complete set of equations (Eq. 4.4). Unlike the approximate equations (Eq. 6.1), where a variety of techniques are available for their analysis, only the MSR method is sufficiently general to handle the additional complexities of the complete set of equations. The complications arise because of the presence of the three fields ρ, $\vec{g}$, and $\vec{V}$. While the MSR method is standard,[52] it is also rather involved. We therefore give a brief review of the basic steps in the formulation.

Let us, for the moment, deal with the general case where we have an equation of motion of the type

$$\frac{\partial \psi(1)}{\partial t_1} = H_1[\psi] + \theta(1), \tag{8.1}$$

where the index 1 labels space, time, and any other index carried by the field ψ. Averages of the form $\langle \psi(1)\psi(2)\ldots\psi(N)\rangle$ correspond to averages over the noise source $\theta(1)$

$$\langle G(\psi)\rangle = \frac{1}{I_0} \int D(\theta)e^{-A_0(\theta)}G(\psi), \tag{8.2}$$

where $G(\psi)$ is some function of ψ, which, via Eq. (8.1), is a function of θ. I_0 is a constant defined such that $\langle 1 \rangle = 1$. The Gaussian nature of the noise requires that $A_0(\theta)$ be of the form

$$A_0(\theta) = \frac{\beta}{4} \int d1 \int d2\, \theta(1)\Gamma^{-1}(1,2)\theta(2). \tag{8.3}$$

Using the identity

$$\int \Gamma(\theta)\frac{\partial}{\partial\theta(1)}[\theta(2)e^{-A_0(\theta)}] = 0, \tag{8.4}$$

and assuming $\Gamma^{-1}(12)$ is symmetric, it is simple to show that the autocorrelation of the noise is given by

$$\langle \theta(1)\theta(2)\rangle = 2k_B T\Gamma(1,2). \tag{8.5}$$

The generating functional for all of the correlation functions of interest is of the form

$$Z_U = I_1 \int D(\theta) e^{-A_0(\theta)} \exp\left[\int d1\, U(1)\psi(1)\right], \qquad (8.6)$$

where I_1 is a normalization constant, and N-point correlation functions are generated from Z_U using

$$\langle \psi(1)\psi(2)\ldots\psi(N)\rangle = \frac{1}{Z_U} \frac{\delta}{\delta U(1)} \frac{\delta}{\delta U(2)} \cdots \frac{\delta}{\delta U(N)} Z_U. \qquad (8.7)$$

Z_U, given by Eq. (8.6), is not convenient for generating the perturbation theory expansion. We can transform it into a more manageable form by using the identity

$$e^{-A_0(\theta)} = I_2 \int D(\hat{\psi}) \exp\left[-\int d1 \int d2[\hat{\psi}(1)\beta^{-1}\Gamma(12)\hat{\psi}(2) + i\theta(1)\hat{\psi}(2)\delta(12)]\right]. \qquad (8.8)$$

The left-hand side is obtained by completing the square in the $\hat{\psi}$ integration and choosing I_2 to cancel the constant coming from the resulting Gaussian integration. We have then

$$Z_U = I_1 I_2 \int D(\theta) \int D(\hat{\psi}) \exp\left[\int d1\, U(1)\right]$$

$$\times \exp\left[-\int d1 \int d2[\hat{\psi}(1)\beta^{-1}\Gamma(1,2)\hat{\psi}(2) + i\theta(1)\delta(1,2)\hat{\psi}(2)]\right]. \quad (8.9)$$

The final step in transforming Z_U is to change variables[53] of integration from θ to ψ. We will use a "causal" continuation from discrete to continuous time, where the Jacobian of the transformation is a constant, and obtain

$$Z_U = I_3 \int D(\psi) \int \Gamma(\hat{\psi}) e^{-A_U[\psi, \hat{\psi}]}, \qquad (8.10)$$

where I_3 is a constant, and

$$A_U[\psi, \hat{\psi}] = \int d1 \int d2\, \hat{\psi}(1)\beta^{-1}\Gamma(1,2)\hat{\psi}(2) + i \int d1\, \theta(1)\hat{\psi}(1) - \int d1\, U(1)\psi(1).$$

$$(8.11)$$

We then replace the noise in Eq. (8.11) in the form

$$\theta(1) = \frac{\partial \psi(1)}{\partial t_1} - H_1[\psi], \tag{8.12}$$

so that we have the "action"

$$A_U[\psi, \hat{\psi}] = \int d1 \int d2\, \hat{\psi}(1)\beta^{-1}\Gamma(1,2)\hat{\psi}(2)$$

$$+ i \int d1\, \hat{\psi}(1)\left[\frac{\partial}{\partial t_1}\psi(1) - H_1[\psi]\right] - \int d1\, U(1)\psi(1). \tag{8.13}$$

We now specialize the development to the fluid equations. We use the notation $1 = (\vec{x}_1, t_1)$ and we assume on the shortest distance scale that the system is set on a lattice. We note first for fluids that the density variable is a bit peculiar since there is no associated dissipative coupling ($\Gamma_{\rho\alpha} = 0$). Therefore, the only appearance of $\hat{\rho}$ in the action is given by

$$i \int d1\, \hat{\rho}(1)\left[\frac{\partial \rho(1)}{\partial t_1} + \nabla \cdot \vec{g}(1)\right], \tag{8.14}$$

and the functional integral over $\hat{\rho}$ reduces to a δ functional enforcing the continuity equation.

Similarly, the relation $\vec{g} = \rho\vec{V}$ is enforced via a δ function constraint. One has then, in Z_U, the product

$$e^{-i\int d1\, \hat{\psi}(1)H_1[\psi]} \int D(\vec{V}) \prod_{\vec{x}} \delta(\vec{g} - \rho\vec{V}).$$

It is very useful to introduce the additional field $\hat{V}_i(x,t)$ in the integral representation for the δ functional $\delta(\vec{g} - \rho\vec{V})$ which generates the term in the action

$$i \int d1 \sum_i \hat{V}_i(1)[g_i(1) - \rho(1)V_i(1)]. \tag{8.15}$$

The action, in the absence of the source term proportional to U, is given then by

$$A[\psi, \hat{\psi}] = \int d1 \left[\sum_{i,j} \hat{g}_i(1) \beta^{-1} \hat{\Gamma}^0_{ij}(\vec{x}_1) \hat{g}_j(1) + i\hat{\rho}(1) \left[\frac{\partial \rho(1)}{\partial t_1} + \nabla_1 \cdot \vec{g}(1) \right] \right.$$

$$+ i \sum_i \hat{g}_i(1) \left[\frac{\partial}{\partial t_1} g_i(1) + \rho(1) \nabla_1^i \frac{\delta F_u}{\delta \rho(1)} + \sum_j \left[\{ \nabla_1^j [\rho(1) V_i(1) V_j(1)] \} \right. \right.$$

$$\left. + \hat{\Gamma}^0_{ij}(\vec{x}_1)(1) V_j(1)] + i \sum_i \hat{V}_i(1) [g_i(1) - \rho(1) V_i(1)] \right] \qquad (8.16)$$

where $\hat{\Gamma}_{ij}(\vec{x}_1)$ is given by Eq. (4.8).

A key reason for changing over to this formalism is that A_U is a polynomial in the ψ and $\hat{\psi}$'s, and standard field-theoretical methods can be used. If we carry out the rescaling

$$\hat{\psi}(1) \to \beta^{-1/2} \hat{\psi}'(1),$$

$$\psi(1) \to \frac{1}{\beta^{-1/2}} \psi'(1),$$

we see that the quadratic components are of $0(1)$ and the higher-order terms are of $0[(k_B T)^{(n-1)/2}]$, where n is the power of ψ in the nonlinear term. Thus we can systematically compute corrections to the Gaussian theory in powers of $k_B T$.

As discussed above, we are interested in the simplest case where F_u is local quadratic form (Eq. 7.6). This choice generates a quadratic term in the action as well as a cubic term

$$i \int d1 \sum_i \hat{g}_i(1) \frac{1}{2} \chi_0^{-1} \nabla_i (\delta\rho)^2.$$

B. Correlation and Response Functions

In developing the theory it is useful to introduce some notation. Let $\psi_\alpha(1)$ be a vector where α runs over $\rho, \hat{\rho}, g_i, \hat{g}_i, V_i$, and $\hat{V}_i$ and one labels space $\vec{x}_1$ and time t_1. We also use the notation $\hat{\alpha}$ to indicate the set including the hatted variables $\hat{\rho}, \hat{g}_i$, and $\hat{V}_i$. The building blocks in the MSR theory are the correlation functions

$$G_{\alpha\beta}(12) = \langle \psi_\alpha(1) \psi_\beta(2) \rangle. \qquad (8.17)$$

We can gain some feeling for the structure of the theory by neglecting the cubic and quartic couplings in the action $A[\psi]$. Denoting quantities in the Gaussian theory by a superscript 0, the correlation functions satisfy the

TABLE I

The inverse of the zeroth-order matrix $G^0_{\alpha\beta}$

	ρ	g_j	V_j	$\hat{\rho}$	$\hat{g}_j$	$\hat{V}$
ρ	0	0	0	$-\omega$	$q_j c_o^0$	0
g_i	0	0	0	q_i	$-\omega\delta_{ij}$	$i\delta_{ij}$
V_i	0	0	0	0	iL_{ij}	$-i\rho_o\delta_{ij}$
$\hat{\rho}$	ω	$-q_j$	0	0	0	0
$\hat{g}_i$	$-q_i c_o^2$	$\omega\delta_{ij}$	iL_{ij}	0	$2\beta^{-1}L_{ij}$	0
$\hat{V}_i$	0	$i\delta_{ij}$	$-i\rho_o\delta_{ij}$	0	0	0

matrix equation

$$\sum_\gamma \int d3 [G^{-1}(1,3)]^0_{\alpha\gamma} G^0_{\gamma\beta}(3,2) = \delta_{\alpha\beta}\delta(1,2). \tag{8.18}$$

After Fourier transformation over space and time, Eq. (8.18) reduces to

$$\sum_\gamma [G^{-1}(q,\omega)]^0_{\alpha\gamma} G^0_{\gamma\beta}(q,\omega) = \delta_{\alpha\beta}, \tag{8.19}$$

where the matrix elements $[G^{-1}(q,\omega)]^0_{\alpha\beta}$ are given in Table I.

The inversion of the matrix $(G^{-1})^0$ to obtain G^0 is facilitated by the realization that the transverse and longitudinal components of $\vec{g}$, $\vec{V}$, and their hatted counterparts can be treated separately. The transverse components $\vec{q}\cdot\vec{q}_T = 0$ do not couple into the density and its hatted conjugate $\hat{\rho}$ and one easily finds the various correlation functions given in Table II.

TABLE II

The transverse part of the zeroth-order matrix $G^0_{\alpha\beta}$.

$W_0 = \rho_0\omega + iq^2\eta_0$

	g_t	V_t	$\hat{g}_t$	$\hat{V}_t$
g_t	$\dfrac{2\beta^{-1}\eta_0 q^2\rho_0^2}{W_0 W_0^*}$	$\dfrac{2\beta^{-1}\eta_0 q^2\rho_0}{W_0 W_0^*}$	$\dfrac{\rho_0}{W_0}$	$\dfrac{\eta_0 q^2}{W_0}$
V_t	$\dfrac{2\beta^{-1}\eta_0 q^2\rho_0}{W_0 W_0^*}$	$\dfrac{2\beta^{-1}\eta_0 q^2}{W_0 W_0^*}$	$\dfrac{1}{W_0}$	$\dfrac{i\omega}{W_0}$
$\hat{g}_t$	$\dfrac{-\rho_0}{W_0^*}$	$\dfrac{-1}{W_0^*}$	0	0
$\hat{V}_t$	$\dfrac{-\eta_0 q^2}{W_0^*}$	$\dfrac{-i\omega}{W_0^*}$	0	0

TABLE III

The longitudinal part of the zeroth-order matrix $G_{\alpha\beta}^0$. $D_0 = \rho_0(\omega^2 - q^2 c_o^2) + i\omega q^2 \Gamma_0$

	ρ	g_l	V_l	$\hat{\rho}$	$\hat{g}_l$	$\hat{V}_l$
ρ	$\dfrac{2\beta^{-1}\Gamma_0 q^4 \rho_0^2}{D_0 D_0^*}$	$\dfrac{2\beta^{-1}\Gamma_0 q^3 \omega \rho_0^2}{D_0 D_0^*}$	$\dfrac{2\beta^{-1}\Gamma_0 q^3 \omega \rho_0}{D_0 D_0^*}$	$\dfrac{\rho_0\omega + i\Gamma_0 q^2}{D_0}$	$\dfrac{q\rho_0}{D_0}$	$\dfrac{\Gamma_0 q^3}{D_0}$
g_l	$\dfrac{2\beta^{-1}\Gamma_0 q^3 \omega \rho_0^2}{D_0 D_0^*}$	$\dfrac{2\beta^{-1}\Gamma_0 q^2 \omega^2 \rho_0^2}{D_0 D_0^*}$	$\dfrac{2\beta^{-1}\Gamma_0 q^2 \omega^2 \rho_0}{D_0 D_0^*}$	$\dfrac{q c_o^2 \rho_0}{D_0}$	$\dfrac{\omega\rho_0}{D_0}$	$\dfrac{\omega\Gamma_0 q^2}{D_0}$
V_l	$\dfrac{2\beta^{-1}\Gamma_0 q^3 \omega \rho_0}{D_0 D_0^*}$	$\dfrac{2\beta^{-1}\Gamma_0 q^2 \omega^2 \rho_0}{D_0 D_0^*}$	$\dfrac{2\beta^{-1}\Gamma_0 q^2 \omega^2}{D_0 D_0^*}$	$\dfrac{q c_o^2}{D_0}$	$\dfrac{\omega}{D_0}$	$\dfrac{i(\omega^2 - q^2 c_o^2)}{D_0}$
$\hat{\rho}$	$\dfrac{-(\rho_0\omega - i\Gamma_0 q^2)}{D_0^*}$	$\dfrac{-q c_o^2 \rho_0}{D_0^*}$	$\dfrac{-q c_o^2}{D_0^*}$	0	0	0
$\hat{g}_l$	$\dfrac{-q\rho_0}{D_0^*}$	$\dfrac{-\omega\rho_0}{D_0^*}$	$\dfrac{-\omega}{D_0^*}$	0	0	0
$\hat{V}_l$	$\dfrac{-\Gamma_0 q^3}{D_0^*}$	$\dfrac{-\omega\Gamma_0 q^2}{D_0^*}$	$\dfrac{i(\omega^2 - q^2 c_o^2)}{D_0^*}$	0	0	0

We note several properties, valid for the G^0, which hold quite generally:

$$G_{\psi_i \psi_j}(\vec{q}, \omega) = 0 \qquad (8.20)$$

and the $G_{\psi_i \psi_j}$ and $G_{\psi_j \psi_i}$ act like, and we shall refer to them as, response functions. Note that they are either retarded or advanced. $G_{\psi_i \psi_j}(\vec{q}, \omega)$ is analytic in the lower half of the complex ω plane.

We also note, for example, that the transverse current correlation function

$$G_{gg}^{T,0} = \frac{2\beta^{-1}\eta_0 q^2}{\omega^2 + (\eta_0 q^2/\rho_0)^2} \qquad (8.21)$$

has the usual hydrodynamical form.[10]

The longitudinal correlation functions can also be worked out explicitly and are given in Table III. The analytic structure is the same as in the transverse case, but the spectrum now involves the traveling waves associated with sound propagation.

C. Nonperturbative Results

Including all of the nonlinear terms in the full action we can generalize Eq. (8.18) to the form

$$\int d\bar{3} \sum_{\gamma} ((G_{\alpha\gamma}^{0-1}(1\bar{3}) - \Sigma_{\alpha\gamma}(1\bar{3}))G_{\gamma\beta}(\bar{3}2)) = \delta_{\alpha\beta}\delta(12), \qquad (8.22)$$

where the self-energies $\Sigma_{\alpha\beta}(12)$ contain the effects of the nonlinearities. In Das and Mazenko it is shown how to perform a perturbation theory expansion for the self-energies. We shall be more concerned here with the general structure of the theory making, only rather weak assumptions about the self-energies.

The theory, which is rather complicated as it stands, can be simplified considerably using general principles.

Using the basic symmetry result $G_{\alpha\beta}(12) = G_{\beta\alpha}(21)$, we can easily show that

$$G_{\alpha\beta}(\vec{q}, \omega) = -G_{\beta\alpha}^*(\vec{q}, \omega). \tag{8.23}$$

It is then clear from the Dyson's equation (Eq. 8.22), defining the self-energies, that

$$\Sigma_{\alpha\beta}(\vec{q}, \omega) = -\Sigma_{\beta\alpha}^*(\vec{q}, \omega). \tag{8.24}$$

Using the continuity equation we can show that

$$\omega G_{\rho\beta}(\vec{q}, \omega) - \vec{q} \cdot G_{\vec{g}\beta}(\vec{q}, \omega) = \delta_{\beta, \hat{\rho}}. \tag{8.25}$$

Thus we have a simple relation between $G_{\rho\beta}$ and the longitudinal part of $G_{g\beta}$.

In a number of nonlinear problems there is available a fluctuation–dissipation theorem (FDT) relating correlation functions $G_{\psi\psi}$ and response functions $G_{\psi\hat{\psi}}$. In the present case we have the restricted fluctuation–dissipation theorem

$$G_{V_i\beta}(\vec{q}, \omega) = -2\beta^{-1}ImG_{\hat{g}_i\beta}(\vec{q}, \omega), \tag{8.26}$$

where β is an unhatted variable. It is straightforward to verify that Eq. (8.26) is satisfied by all of the zeroth-order quantities in Table II. Using Eq. (8.26) we can relate a number of the response functions and correlation functions.

D. Transverse Self-Energies

The full matrix G can be divided into its longitudinal and transverse parts as in the linear case. In a similar fashion the self-energy $\Sigma_{\alpha_i\beta_j}(\vec{q}, \omega)$, where α_i and β_j are in the set of vectors $(g_i, \hat{g}_i, V_i, \hat{V}_i)$ and i and j are vector labels, can be written in the form

$$\Sigma_{\alpha_i\beta_j}(\vec{q}, \omega) = \hat{q}_i\hat{q}_j\Sigma_{\alpha\beta}^L(\vec{q}, \omega) + (\delta_{ij} - \hat{q}_i\hat{q}_j)\Sigma_{\alpha\beta}^T(\vec{q}, \omega). \tag{8.27}$$

Since the transverse case contains only the components of the vector fields, it is less complicated and we will consider it first.

One significant simplification of the problem is that none of the cubic or quartic terms in the action contain a $\vec{g}$ field. Thus all self-energies with a g_i index vanish,

$$\Sigma_{g_i\alpha}(\vec{q}, \omega) = \Sigma_{\alpha g_i}(\vec{q}, \omega) = 0. \tag{8.28}$$

Simplification also occurs in the long-wavelength limit. This results from the conservation of momentum which says that every external $\hat{g}_i$ vertex contributing to $\Sigma^T_{\hat{g}_i\beta}$ supplies a factor q_i. Since the system is isotropic we find that $\Sigma^T_{\hat{g}\beta}(\vec{q}, \omega)$ must be of $0(q^2)$.

Using the simplifications discussed above, we can invert Eq. (8.22) to obtain the response functions in the form

$$G^T_{\alpha\beta}(\vec{q}, \omega) = \frac{M_{\alpha\beta}}{W}, \tag{8.29}$$

where the $M_{\alpha\beta}$ are given by

$$M_{g\hat{g}} = 1, \tag{8.30a}$$

$$M_{g\hat{V}} = q^2\eta(\vec{q}, \omega), \tag{8.30b}$$

$$M_{V\hat{g}} = \frac{1}{\rho_0} \frac{1}{W_V(\vec{q}, \omega)}, \tag{8.30c}$$

$$M_{V\hat{V}} = \frac{i\omega}{\rho_0} \frac{1}{W_V(\vec{q}, \omega)}, \tag{8.30d}$$

where

$$W = \omega + iq^2\eta(\vec{q}, \omega), \tag{8.31}$$

$$W_V = 1 - i\Sigma^T_{\hat{V}V}(\vec{q}, \omega)/\rho_0, \tag{8.32}$$

and

$$q^2\eta(\vec{q}, \omega) \equiv [q^2\eta_0 + i\Sigma^T_{\hat{g}V}(\vec{q}, \omega)]/\rho_0 W_V(\vec{q}, \omega). \tag{8.33}$$

Using the fluctuation–dissipation theorem (Eq. 8.26), we find the fluctuation functions, which are the Laplace transform of the correlation functions,

$$C^T_{V\beta}(\vec{q}, \omega) = \beta^{-1}G_{\beta\hat{g}}(\vec{q}, \omega) \tag{8.34}$$

and

$$C_{VV}^T(\vec{q}, \omega) = \frac{\beta^{-1}}{\rho_0 \, W_V(\vec{q}, \omega)} \frac{1}{\omega + iq^2 \eta(\vec{q}, \omega)}, \tag{8.35}$$

$$C_{V_g}^T(\vec{q}, \omega) = \frac{1}{\omega + iq^2 \eta(\vec{q}, \omega)}. \tag{8.36}$$

We can gain some feeling for the physics of the situation by concentrating on $C_{VV}^T(\vec{q}, \omega)$. We first look at the static behavior. We have

$$\lim_{\omega \to \infty} \omega \, C_{VV}^T(\vec{q}, \omega) = \int \frac{d\omega}{2\pi} G_{VV}^T(\vec{q}, \omega), \tag{8.37}$$

$$= \chi_{VV}^T(\vec{q}) = \frac{\beta^{-1}}{\rho_0 \, W_V(\vec{q}, \infty)}. \tag{8.38}$$

Assuming that the self-energy $\Sigma_{VV}^T(\vec{q}, \omega)/\rho_0$ vanishes as $1/\omega$ for large ω, $W_V(\vec{q}, \infty) = 1$ and

$$\chi_{VV}^T(\vec{q}) = \frac{\beta^{-1}}{\rho_0}, \tag{8.39}$$

as one would expect. The equilibrium density is unchanged from its linearized value. Similarly,

$$\chi_{V_g}^T(\vec{q}) = \beta^{-1}. \tag{8.40}$$

Let us note that in the linearized theory $W_V = 1, \eta = \eta_0$, and C_{VV}^T has the usual spectrum for a conserved variable. In the nonlinear system it is instructive to look at the long wavelength limit where

$$C_{VV}^T(0, \omega) = \frac{\beta^{-1}}{\omega \rho_0 \, W_V(\vec{0}, \omega)}. \tag{8.41}$$

We can show exactly, using the FDT, that

$$\Sigma_{VV}^{T'}(\vec{0}, \omega) = \frac{\beta\omega}{2} \Sigma_{VV}^{T'}(\vec{0}, \omega), \tag{8.42}$$

and we can write

$$W_V(\vec{0}, \omega) = W_0(\omega)[1 + i\omega\tau(\omega)], \tag{8.43}$$

where

$$W_0(\omega) = 1 + \Sigma_{\hat{V}V}^{T''}(\vec{0}, \omega)/\rho_0, \tag{8.44}$$

and

$$\tau(\omega) = -\frac{\beta}{2\rho_0}\Sigma_{\hat{V}\hat{V}}^{T'}(\vec{0}, \omega)/W_0(\omega). \tag{8.45}$$

Then

$$G_{VV}(\vec{0}, \omega) = -2Im\frac{\beta^{-1}}{\omega\rho_0 W_0}[1 + i\omega\tau(\omega)]^{-1},$$

$$= \frac{2\pi\beta^{-1}}{\rho_0}\frac{\delta(\omega)}{W_0(o)} + \frac{2\beta^{-1}}{\rho_0 W_0(\omega)}\frac{\tau(\omega)}{[1 + (\omega\tau(\omega))^2]}, \tag{8.46}$$

and there is a "conserved" portion proportional to $\delta(\omega)$ and a relaxational component with a relaxation time $\tau(\omega)$.

Clearly $C_{V_g}^T(\vec{q}, \omega)$ is purely hydrodynamic in the long-wavelength limit. The general expression for $G_{gg}^T(\vec{q}, \omega)$ is very complicated, but in the long-wavelength limit we obtain

$$G_{gg}^T(\vec{q}, \omega) = \frac{2\beta^{-1}}{WW^*}q^2\eta'(\vec{0}, \omega)\rho_0 W_V'(\vec{0}, \omega). \tag{8.47}$$

E. Longitudinal Case

In the longitudinal case we follow a procedure similar to that in the transverse case. However, this case is more complicated since we have three fields $(\rho, \vec{g}, \vec{V})$ in both the hatted and unhatted sets. Inverting the G^{-1} matrix in Eq. (8.22) we obtain the response functions in the form

$$G_{\alpha\hat{\beta}}^L(\vec{q}, \omega) = \frac{N_{\alpha\hat{\beta}}}{D} \tag{8.48}$$

with the matrix N given in Table IV, and

TABLE IV
The matrix $N_{\alpha\hat{\beta}}$ determining the response function $G_{\alpha\hat{\beta}}$ in Eq. (8.48)

	$\hat{\rho}$	$\hat{g}$	$\hat{V}$
ρ	$\omega + iq^2 D_L$	q	$q^3 D_L$
g	$qc^2 + L\Sigma_{\hat{V}\rho}$	ω	ωL
V	$(qc^2 + i\omega\Sigma_{\hat{V}\rho})/\rho_o W_L$	$(\omega + iq\Sigma_{\hat{V}\rho})/\rho_o W_L$	$i(\omega^2 - q^2c^2)/\rho_o W_L$

$$D = \omega^2 - c^2 q^2 + iq^2 D_L(\omega + iq\Sigma_{\hat{V}\rho}^T), \tag{8.49}$$

$$q^2 D_L(\vec{q}, \omega) = [\vec{q}^2 D_0 + i\Sigma_{\hat{g}V}^L(\vec{q}, \omega)]/(\rho_0 W_L(\vec{q}, \omega)), \tag{8.50}$$

$$qc^2(\vec{q}, \omega) = qc_0^2 + \Sigma_{\hat{g}\rho}^L(\vec{q}, \omega), \tag{8.51}$$

$$\rho_0 W_L(\vec{q}, \omega) = \rho_0 - i\Sigma_{\hat{V}V}^L(\vec{q}, \omega). \tag{8.52}$$

Similarly,

$$G_{\beta\alpha}^L = \frac{N_{\beta\alpha}}{-D^*}, \tag{8.53}$$

where

$$N_{\alpha\beta} = (N_{\alpha\beta})^*. \tag{8.54}$$

The correlation functions are given in this case by

$$G_{\alpha\beta}^L = -\sum_{\gamma, \delta} G_{\alpha\gamma}^L C_{\gamma\delta}^L G_{\delta\beta}^L, \tag{8.55}$$

where the summations are over ρ, g, and V. C^L is given by

$$C_{\hat{\alpha}\beta\alpha}^L = -\Sigma_{\hat{\alpha}\beta\alpha}^L + \delta_{\hat{\alpha}, \hat{g}}\delta_{\hat{\beta}, \hat{g}}2\beta^{-1}q^2\Gamma_0 \tag{8.56}$$

and is zero if $\hat{\alpha}$ or $\hat{\beta}$ equals $\hat{\rho}$. C^L has only four nonzero elements. Of these components $\Sigma_{\hat{g}\hat{g}}^L$ and $\Sigma_{\hat{V}\hat{V}}^L$ are real, and $\Sigma_{\hat{g}\hat{V}}^L$ is imaginary. The FDT can be used to express $G_{V\rho}$, G_{Vg}, and G_{VV} simply in terms of $G_{\hat{g}\rho}$, $G_{\hat{g}g}$, and $G_{\hat{g}V}$, respectively.

As in the transverse case, the analysis is simplified by looking at the long-wavelength small-frequency limit where we can write

$$\Sigma_{\hat{g}V}^L = -iq^2\gamma_{\hat{g}V}, \tag{8.57a}$$

$$\Sigma_{\rho\hat{g}}^L = q\gamma_{\rho\hat{g}}, \tag{8.57b}$$

$$\Sigma_{\rho\hat{V}}^L = q\gamma_{\rho\hat{V}} \equiv q\gamma, \tag{8.57c}$$

$$\Sigma_{\hat{g}\hat{g}}^L = -q^2\gamma_{\hat{g}\hat{g}}, \tag{8.57d}$$

and the γ's are defined to be real positive numbers. In this case the physical speed of sound is given by

$$c^2 = c_0^2 - \gamma_{\rho\hat{g}}, \tag{8.58}$$

the physical longitudinal viscosity is

$$D_L = \left(D_0 + \frac{\beta}{2}\gamma_{\theta\theta}\right)/\rho_0 W_L \tag{8.59}$$

and, for example,

$$G_{\rho\hat{\rho}}(\vec{q}, \omega) = \frac{\omega + iq^2 D_L}{\omega^2 - q^2 c^2 + iq^2 D_L(\omega + iq^2\gamma)} \tag{8.60}$$

and

$$G_{\rho\rho}(\vec{q}, \omega) = -2\beta^{-1}\chi_{\rho\rho}(\vec{q})G_{\hat{\rho}\rho}(\vec{q}, \omega). \tag{8.61}$$

We can then show that the associated fluctuation function is given by

$$C_{\rho\rho}(\vec{q}, \omega) = \frac{\chi_{\rho\rho}(\vec{q})(\omega + iq^2 D_L)}{\omega^2 - c^2 q^2 + iq^2 D_L(\omega + iq^2\gamma)} \tag{8.62}$$

F. Cutoff for the Leutheusser Transition

The key point we want to make is that the basic structure of $C_{\rho\rho}(\vec{q}, \omega)$ differs significantly from that given by Eq. (3.18)

$$C_{\rho\rho}(\vec{q}, \omega) = \frac{\chi_{\rho\rho}(\vec{q})(\omega + iq^2 D_L)}{\omega^2 - c^2 q^2 + iq^2\omega D_L}. \tag{8.63}$$

In the hydrodynamic limit where the main contribution to $C_{\rho\rho}(\vec{q}, \omega)$ comes from the regime $\omega \sim cq$, the term proportional to γq^2 can be dropped relative to ω in Eq. (8.62). Thus this term $iq^2\gamma$ in the denominator of Eq. (8.62) represents a nonhydrodynamical correction. For our purposes here, however, it is crucial. Suppose, as in the Leutheusser mechanism, that D_L becomes large. In our previous analysis we obtained

$$C_{\rho\rho}(\vec{q}, \omega) \to \frac{\chi_{\rho\rho}(\vec{q})}{\omega}. \tag{8.64a}$$

Now, however, we find

$$C_{\rho\rho}(\vec{q}, \omega) \to \frac{\chi_{\rho\rho}(\vec{q})}{\omega + iq^2\gamma}. \tag{8.64b}$$

Thus we immediately see that there is a mechanism available to cutoff the Leutheusser transition. Instead of a transition to a nonergodic phase, we find, even for very large D_L, that the density fluctuations decay to zero with a time $\tau \sim (q^2\gamma)^{-1}$. The occurrence of the nonhydrodynamic correction proportional to γq^2 arises due to the inclusion of the slow variable $\vec{V}$ into our set of hydrodynamic variables via the nonlinear constraint $\vec{g} = \rho\vec{V}$.

G. Results At First Order in Perturbation Theory

At this stage we only know how to determine the self-energies using perturbation theory. Working at lowest $[0(k_B T)]$ order we see that the picture developed above is substantiated in detail. We obtain the following:

1. The main density feedback mechanism contributing to $D_L(\vec{q}, \omega)$ is that given by Eq. (6.3) which generates the Leutheusser mechanism.

2. The speed of sound is not changed from its zeroth order value at this order.

3. The explicit expression for the damping γ is:

$$\gamma(\vec{q}, \omega) = \frac{c^2\beta}{3\rho^3} \int_0^\infty dt\, e^{i\omega t} \int \frac{d^3k}{(2\pi)^3} \frac{1}{k^2} \left(\frac{d}{dt} G_{\rho\rho}(\vec{k}, t)\right)\left(\frac{d}{dt} G_{\rho\rho}(\vec{q} - \vec{k}, t)\right)$$
$$+ \frac{c^2\beta}{3\rho^3} \int_0^\infty dt\, e^{i\omega t} \int \frac{d^3k}{(2\pi)^3} G_{\rho\rho}(\vec{k}, t) G_{VV}^T(\vec{q} - \vec{k}, t). \qquad (8.65)$$

Thus we have a set of nonlinear equations for D_L, γ, and $C_{\rho\rho}$ which appear to give the dominant contribution in the highest density and lowest temperature limits.

H. Numerical Analysis

We can gain some feeling for the solution of this coupled set of equations (8.62), (6.3), and (8.65) by assuming, as in Section VI, that $C_{\rho\rho}(\vec{q}, t)$ is insensitive to wavenumber and defining

$$\phi(t) = C_{\rho\rho}(\vec{q}, t). \qquad (8.66)$$

In this case the equations for the longitudinal viscosity and the damping γ take the form

$$D(\omega) = D_0 + \lambda c^2 \int_0^\infty e^{i\omega t}\phi^2(t)\, dt, \qquad (8.67)$$

$$\gamma(\omega) = \frac{\lambda}{3} \int dt\, e^{i\omega t} [\dot{\phi}(t)]^2 \equiv \frac{\lambda}{3} I(\omega), \qquad (8.68)$$

where the coupling λ is defined by

$$\lambda = \frac{(k_B T)\Lambda^3}{6\pi^2 \rho c^2},\qquad (8.69)$$

where, for simplicity, we have dropped the term contributing to $\gamma(\omega)$ which depends on C_{VV}^T. This term gives a simple additive contribution to γ which does not have a strong frequency dependence. With these approximations we can derive a single integro-differential equation for $\phi(t)$:

$$\ddot{\phi}(t) + q^2 \lambda_1 \left[\dot{\phi}(t) + \lambda \int_0^t d\tau \phi^2(\tau)\dot{\phi}(t-\tau) \right.$$

$$\left. + \frac{q^2 \lambda}{3\Lambda^2} \int_0^t d\tau \dot{\phi}^2(t-\tau)\left[\phi(\tau) + \lambda \int_0^\tau d\tau' \phi^2(\tau-\tau')\phi(\tau') \right] \right] = 0,\quad (8.70)$$

where

$$\lambda_1 = \frac{\Lambda^2 \Gamma_0}{c^2}.\qquad (8.71)$$

This equation should be compared to the Leutheusser equation obtained in the absence of the γ term,

$$\ddot{\phi}(t) + \Gamma_0 q^2 \dot{\phi}(t) + \lambda q^2 \int_0^t \phi^2(\tau)\dot{\phi}(t-\tau)\,d\tau = 0.\qquad (8.72)$$

One can solve these equations numerically. Notice that they are dependent parametrically on q^2 (measured in units of the cutoff). However, the solution to Eq. (8.72) in the long-time limit is independent of q^2 as long as $q^2 \neq 0$. We show $\phi(t)$, resulting from Eq. (8.72), in Fig. 5 (in dotted lines) for various choices of λ near the transition. For $\lambda = 4$ the viscosity diverges. For $\lambda > 4$ the system is nonergodic and $\phi(t)$ does not decay to zero for long times.

We have solved Eq. (8.70) numerically using the same choice for $\lambda_1 \ (= 1)$ used by Leutheusser, and for which we know there is a transition for $\lambda = 4$. In Fig. 5 we show $\phi(t)$ versus t for several choices of λ. This should be compared with the case where γ correction is ignored (shown in dotted lines). In this figure we see that the system continues to decay even for $\lambda > 4$. A detailed analysis of the data shows that there is exponential decay at sufficiently long time and the decay rate, σ, defined by

$$\sigma = -\dot{\phi}/\phi\qquad (8.73)$$

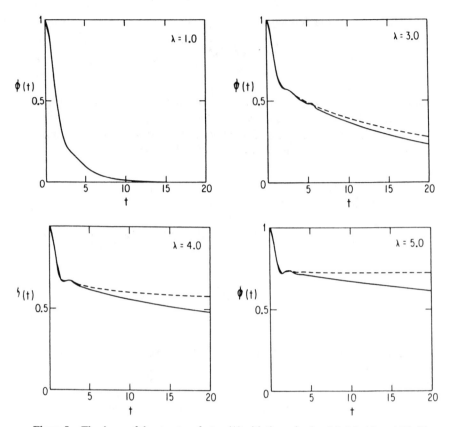

Figure 5. The decay of the structure factor $\phi(t)$ with time t for $\lambda = 1.0$, 3.0, 4.0, and 5.0. The solid lines indicate the results of the numerical integration of Eq. (8.70). The dotted line corresponds to the case where the γ correction is ignored. The value of q/Λ used is $\frac{1}{8}$ (from ref. 4).

is given in Table V as a function of λ. We also give there $D_L(0)$ and $\gamma(0)$ as functions of λ. One finds the type of behavior described above. Looking in particular at D_L, we see that there are two types of behavior. For larger λ one finds that D_L is essentially linear with λ:

$$D_L(0) = -308.69 + 92.76\lambda, \qquad (8.74)$$

and the fit is excellent for $\lambda > 3.75$. Notice that if we extrapolate this result to small λ, one finds that D_L vanishes at $\lambda = 3.33$. For small λ, motivated by the previous theoretical work and the experiments of Taborak et al.,[48] we have

TABLE V
The longitudinal viscosity Γ, I, and σ as a function of λ
resulting from a numerical integration of Eq. (8.70). The value
of q/Λ was $\frac{1}{8}$

λ	σ	I	Γ/Γ_0
0		0.56	1.0
1	0.4054	0.292	2.0
2	0.1434	0.190	4.4
2.5	0.0837	0.158	7.4
3	0.0443	0.133	13.8
3.5	0.2247	0.114	29.6
3.75	0.0166	0.106	44.0
4	0.0132	0.099	62.9
4.25	0.0113	0.093	84.2
4.5	0.0106	0.088	103.8
5	0.0090	0.079	152.1
6	0.0076	0.067	248.4
7	0.0071	0.058	342.6
8	0.0067	0.052	435.8
9	0.0065	0.047	526.6
10	0.0064	0.043	616.7

plotted $D_L^{-1/2}(0)$ versus λ and find a reasonable fit for $1 < \lambda < 4$:

$$D_L^{-1/2}(0) = 0.874 - 0.195\lambda.$$

Notice that this result predicts a transition for $\lambda = 4.48$. Thus there is a remnant of the Leutheusser transition for this choice of parameters. Analysis of the $I(0)$ shows that this behaves like $0.364\lambda^{-0.93}$ for $2 < \lambda < 10$. Since γ is λ times $I(0)$, we see that γ increases slowly as λ increases.

The conclusion we draw is that density fluctuations are effective in increasing the viscosity, but they do not, by themselves, appear sufficient to give the strong temperature dependences seen in the more strongly coupled glassy systems.[54] These conclusions must be taken with the following reservations. (1) We have treated the wavenumber dependences of this system in a very crude and quantitatively unsatisfactory way. As emphasized by Kirkpatrick,[51] the slowing down for dense fluids should be correlated with the peak in the structure factor, which should play an important role in the analysis. (2) Our identification of the parameter λ with increasing density and lower temperature is not credible unless we appeal to the correlations mentioned in (1). In his thesis[5] Das has removed these restrictions associated with the wavenumber and treated a more realistic wavenumber dependence for the structure factor and carried out the wavenumber integrations in the mode-coupling integrals for $D_L(\vec{q}, \omega)$ and $\gamma(\vec{q}, \omega)$ explicitly. He found that the basic physical picture described above still holds and a direct comparison for the shear viscosity with computer simulation data gave qualitative agreement.

IX. CONCLUSION

At this point it appears that the simple equations of fluctuating nonlinear hydrodynamics Eq. (4.4) do not lead to a sharp glass transition. They do lead to a substantial increase of the viscosities as the density is increased and the intermediate (prehydrodynamic) time regime is influenced by a nonhydrodynamic "diffusive" mode. The associated theory shows strong remnants of the Leutheusser transition and the overall behavior looks very similar to that found in experiments and simulations well away from the glass transition.

A final important question is whether this type of approach has anything to say about *the* glass transition. Our conclusions here are that equations of the type given by Eq. (4.4) with simple effective Hamiltonians, such as those studied by Das, will not lead to a glass transition. It is our belief that in order to make contact with the freezing associated with the glass transition, one must generalize the models studied to allow for the underlying thermodynamic liquid–solid transition. This is the direction in which our future research in this area will focus.

Acknowledgments

We thank Dr. Sidney Nagel for useful discussions. We also thank Dr. John Marko for reading the manuscript and making valuable comments. This work was supported by the National Science Foundation Grant No. DMR 85-12901 and by the National Science Foundation Materials Research Laboratory at the University of Chicago.

References

1. E. Leutheusser, *Phys. Rev. A* **29**, 2765 (1984).
2. A rather complete analysis of the Leutheusser transition from a kinetic theory point of view is given in U. Bengtzelius, W. Götze, and A. Sjölander, *J. Phys. C* **17**, 5915 (1984); W. Götze, *Z. Phys. B* **56**, 139 (1984); **60**, 195 (1985). U. Bengtzelius and L. Sjögren, *J. Chem. Phys.* **84**, 1744 (1986). H. DeRaedt and W. Götze, *J. Phys.* **C19**, 2607 (1986). U. Bengtzelius, *Phys. Rev. A* **33**, 3433 (1986). W. Götze and L. Sjögren, *Z. Phys. B* **65**, 415 (1987).
3. S. Das, G. Mazenko, S. Ramaswamy, and J. Toner, *Phys. Rev. Lett.* **54**, 118 (1985).
4. S. Das and G. Mazenko, *Phys. Rev. A* **34**, 2265 (1986).
5. S. Das, *Phys. Rev. A* **36**, 211 (1987).
6. J. C. Maxwell, *The Scientific Papers of J. C. Maxwell*, Dover, New York, 1965.
7. L. Boltzmann, *Lectures on Gas Theory* (S. Brush, trans.), University of California Press, Berkeley, California, 1964.
8. S. Chapman and T. G. Cowling, *The Mathematical Theory of Non-Uniform Gases*, 3rd. ed., Cambridge University Press, London, 1970. Also J. O. Hirschfelder, C. F. Curtiss, and R. B. Bird, *Molecular Theory of Gases and Liquids*, Wiley, New York, 1954.

9. L. P. Kadanoff and P. C. Martin, *Ann. Phys.* (*N.Y.*) **24**, 419 (1963); P. C. Martin in *Many Body Physics*, C. DeWitt and R. Balian (Eds.), Gordon and Breach, New York, 1968.

10. D. Forster, *Hydrodynamic Fluctuations, Broken Symmetry, and Correlation Functions*, Benjamin, Reading, MA, 1975. B. J. Berne and R. Pecora, *Dynamic Light Scattering*, Wiley, New York, 1976. J. P. Boon and S. Yip, *Molecular Hydrodynamics*, McGraw-Hill, New York, 1980. J. P. Hansen and I. R. McDonald, *Theory of Simple Liquids*, 2nd ed., Academic, New York, 1986.

11. M. S. Green, *J. Chem. Phys.* **20**, 1281 (1952); *J. Chem. Phys.* **22**, 398 (1954).

12. R. Kubo, *J. Phys. Soc. Japan*, **12**, 570 (1957).

13. R. W. Zwanzig, *Phys. Rev.* **129**, 486 (1963).

14. B. J. Alder and T. E. Wainwright, *Phys. Rev. Lett.* **18**, 968 (1967).

15. K. Kawasaki and I. Oppenheim, *Phys. Rev.* **136**, A1519 (1964); **139**, A649 (1965); **139**, A1753 (1965); in *Statistical Mechanics*, T. Bak. (Ed.), Benjamin, New York, 1967. *J. Stecki, Phys. Lett.* **19**, 123 (1965). L. K. Haynes, J. R. Dorfman, and M. H. Ernst, *Phys. Rev.* **144**, 207 (1966). J. R. Dorfman and E. G. D. Cohen, *Phys, Lett.* **16**, 124 (1965). J. V. Sengers, *Phys. Rev. Lett.* **15**, 515 (1965). J. Weinstock, *Phys. Rev.* **140**, A461 (1965).

16. For a review see M. H. Ernst, L. K. Haines, and J. R. Dorfman, *Rev. Mod. Phys.* **41**, 296 (1969).

17. Y. Pomeau and P. Resibois, *Phys. Rep.* **19c**, 63 (1975).

18. P. Resibois and M. De Leener, *Classical Kinetic Theory*, Wiley, New York, 1977.

19. B. J. Alder and T. E. Wainwright, *Phys. Rev. A* **1**, 18 (1970); *J. Phys. Soc. Japan* **26**, Supplement (1969); B. J. Alder, D. M. Gass, and T. E. Wainwright, *J. Chem. Phys.* **53**, 3813 (1970); *Phys. Rev. A* **4**, 233 (1971).

20. G. F. Mazenko, *Phys. Rev. A* **9**, 360 (1974).

21. P. M. Furtado, G. F. Mazenko, and S. Yip, *Phys. Rev. A* **14**, 869 (1976); G. F. Mazenko and S. Yip in *Statistical Mechanics, Part B: Time-Dependent Processes* B. J. Berne (Ed.), Plenum Press, New York, 1977.

22. L. Sjögren and A. Sjölander, *Ann. Phys.* (*N.Y.*) **110**, 122, 411 (1978).

23. J. R. Dorfman and E. G. D. Cohen, *Phys. Rev. Lett.* **1257**, (1970), *Phys. Rev. A* **6**, 776 (1972). J. R. Dorfman and E. G. D. Cohen, *Phys, Rev. A* **12**, 292 (1975).

24. P. Resibois, *J. Stat. Phys.* **13**, 393 (1975).

25. A. Einstein, *Ann. Phys.* **17**, 549 (1905); also in *Investigations on the Theory of the Brownian Movement*, Dover, New York, 1956, Chapter 5.

26. P. Langevin, *C.R. Acad. Sci.* (*Paris*) **146**, 530 (1908).

27. A. D. Fokker, *Ann. Phys.* **43**, 812 (1914).

28. M. Planck, *Ann. Phys.* **26**, 1 (1908).

29. M. V. Smoluchowski, *Ann. Phys.* **21**, 756 (1906); **25**, 205 (1908).

30. Y. Nambu, *Phys. Rev. Lett.* **4**, 380 (1960); Y. Nambu and G. Jona-Lasinio, *Phys. Rev.* **122**, 345 (1961); J. Goldstone, *Nuovo Cimento* **19**, 154 (1961); J. Goldstone, A. Salam, and S. Weinberg, *Phys. Rev.* **127**, 956 (1962).

31. R. Zwanzig, in *Lectures in Theoretical Physics*, Vol. 3, Interscience, New York, 1961.

32. H. Mori, *Prog. Theor. Phys.* (*Kyoto*) **34**, 423 (1965).

33. G. Mazenko, in *Correlation Functions and Quasiparticle Interactions in Condensed Matter*, J. W. Halley (Ed.), Plenum, New York, 1978.

34. J. Villain, *Solid State Commun.* **8**, 31 (1970); G. Mazenko, M. Nolan, and R. Freedman, *Phys. Rev. B* **18**, 2281 (1978); P. Horn, L. Corliss, and J. Hastings, *Phys. Rev. Lett.* **40**, 126 (1978).

35. P. C. Martin, O. Parodi, and R. S. Pershan, *Phys. Rev. A* **6**, 2401 (1972), G. Grinstein and R. A. Pelcovits, *Phys. Rev. Lett.* **47**, 856 (1981); *Phys. Rev. A* **26**, 915 (1982), G. Mazenko, R. Ramaswamy, and J. Toner, *Phys. Rev. Lett.* **49**, 51 (1982); *Phys. Rev. A* **28**, 1618 (1983), S. Bhattacharya and J. B. Ketterson, *Phys. Rev. Lett.* **42**, 997 (1982), J. L. Gallani and P. Martinoty, *Phys. Rev. Lett.* **54**, 333 (1985).

170   BONGSOO KIM and GENE F. MAZENKO

36. L. D. Landau and E. M. Lifshitz, *Fluid Mechanics*, Addison Wesley, Reading, MA, 1959.
37. H. Mori and H. Fujisaka, *Prog. Theor. Phys.* **49**, 764 (1973), H. Mori, H. Fujisaka, and H. Shigematsu, *Prog. Theor. Phys.* **51**, 109 (1974).
38. S. Ma and G. F. Mazenko, *Phys. Rev. B* **11**, 4077 (1975).
39. For a semimicroscopic derivation see J. Langer and L. Turski, *Phys. Rev. A* **8**, 3230 (1973).
40. D. Forster, D. R. Nelson, and M. Stephen, *Phys. Rev. A* **16**, 732 (1977). C. DeDominicis and P. C. Martin, *Phys. Rev. A* **19**, 419 (1979). V. Yakhot, *Phys. Rev. A* **23**, 1486 (1981). J. P. Fournier and U. Frisch, *Phys. Rev. A* **17**, 747 (1978); **28**, 1000 (1983).
41. P. C. Hohenberg and B. I. Halperin, *Rev. Mod. Phys.* **49**, 435 (1977).
42. P. C. Martin, E. D. Sigga, and H. A. Rose, *Phys. Rev. A* **8**, 423 (1973).
43. E. Leutheusser, *J. Phys. C* **15**, 2801, 2827 (1982).
44. H. Vogel, *Phys. Z.* **22**, 645 (1921).
45. G. S. Fulcher, *J. Am. Ceram. Soc.* **8**, 339 (1925).
46. J. Wong and C. A. Angell, *Glass: Structure by Spectroscopy*, Dekker, New York, 1976, Chapter 11.
47. N. O. Birge and S. R. Nagel, *Phys. Rev. Lett.* **54**, 2674 (1985); Y. Jeong, S. R. Nagel, and S. Bhattacharya, *Phys. Rev. A* **34**, 602 (1986). Also in Ann. N.Y. Acad. Sci., C. Angel and M. Goldstein, Eds., *Dynamic Aspects of Structural Change in Liquids and Glasses*, Vol. 484, New York, 1986.
48. P. Taborek, R. N. Kleinman, and D. J. Bishop, *Phys. Rev. B.* **34**, 1835 (1986).
49. C. A. Angell, J. Clarke, and L. V. Woodcock, *Adv. Chem. Phys.* **48**, 397 (1981); C. A. Angell, *Ann. N.Y. Acad. Sci.* **371**, 136 (1981); J. R. Fox and H. C. Andersen, *J. Phys. Chem.* **88**, 4019 (1984); J. J. Ullo and S. Yip, *Phys. Rev. Lett.* **54**, 1509 (1985).
50. E. D. Siggia, *Phys. Rev. A* **32**, 3135 (1985).
51. T. R. Kirkpatrick, *Phys. Rev. A* **31**, 939 (1985).
52. U. Deker and F. Haake, *Phys. Rev. A* **11**, 2043 (1975); **12**, 1629 (1975). R. Bausch, H. Jensen, and H. Wagner, *Z. Phys. B* **24**, 113 (1976). H. Jenssen, in *Dynamical Critical Phenomena and Related Topics*, C. P. Enz (Ed.), Springer-Verlag, New York, 1979, U. Deker, *Phys. Rev. A* **19**, 846 (1979). R. Jensen, *J. Stat. Phys.* **25**, 183 (1981).
53. C. De Dominicis and L. Peliti, *Phys. Rev. B* **18**, 353 (1978).
54. C. A. Angell in *Relaxations in Complex Systems* K. Ngai and G. Wright (Ed.), U.S. Government Printing Office, Washington D.C., 1984.

STRUCTURE OF THE ELECTRIC DOUBLE LAYER

L. BLUM

Department of Physics, University of Puerto Rico, Rio Piedras, Puerto Rico

CONTENTS

INTRODUCTION: THE EXPERIMENTAL EVIDENCE

The general problem of the interface between two phases which are charged and or conducting is of relevance to a number of systems which occur in nature: colloids, micelles, membranes, solid–solution interfaces in general, and

171

metal–solution interfaces in particular, form a bewildering array of systems of enormous complexity. The investigation of the structure of these systems pose considerable difficulties, both experimentally as well as theoretically. The experimental problem is that the interface has 10^{-8} particles relative to the bulk, solid, or liquid phases. For this reason one needs a surface specific method, which is able to discriminate between the signal from the surface and the rest. Electrons do not penetrate into solids and for that reason have been used extensively for the determination of the surface structure of solids. However, they must be used in vacuum and that precludes their use in the study of the liquid–solid interface. The study of electrode surfaces removed from the liquid cell under various conditions has provided an enormous wealth of useful data which we will not try to review here.[1] The only way to understand the relation between the ex situ and in situ systems is to measure both.

Neutrons are another possible choice: unfortunately there is no mechanism by which the neutrons can be made surface selective, and for that reason only systems with large specific surface, which in general are disordered, can be studied.

X rays are a viable alternative for the study of solid–liquid interfaces. The X-ray probes are reviewed in the excellent chapter by H. D. Abruna. There are three techniques that have been used to determine structural features in well-characterized metal electrolyte interfaces:

1. EXAFS or extended X-ray absorption fine structure[2] which permits the determination of the near-neighbor structure of a given target atom and also yields information about the electronic state of that atom when it is adsorbed at the surface.[3]

2. GIXS, or grazing incidence X-ray diffraction.[4] This method permits the determination of the in-plane structure of an adsorbed monolayer of a target atom on the surface. It is a very exacting technique, and requires that the structure of the adsorbed monolayer should be different from that of the substrate.[5]

3. Standing wave methods.[6] In this case an X-ray standing wave is set up at the interface of the solid and the fluid. There are several modes in which these standing waves can be formed; they allow the determination of the distance from the surface of the solid into the fluid phase.[7]

The spectroscopic methods using ultraviolet, visible, or Raman spectroscopy are very useful in in situ probes.[8] Although they are not directly related to the geometrical structural parameters, these can be extracted from theoretical considerations with a fair degree of reliability. The optical spectroscopic methods do not require special installations such as the synchrotron and are most useful for complex molecular species. The techniques are the surface-enhanced Raman,[9] surface infrared spectroscopy,[10] and second harmonic generation,[11] which permits discrimination between different geometries of the adsorbates on single crystal surfaces.

A technique that provides direct structural information of the electrode surface is the Scanning Tunnel Microscope (STM). It has been recently shown by various groups that the STM is capable of resolving structural details of metal surfaces in contact with electrolytic solutions.[12] However, when the electrochemical potential is scanned, the tunnel voltage of the STM also changes. This does not affect the study of surface geometry, since the images are relatively independent of the tunnel voltage.[13] The resolution of the STM pictures of the metal electrolyte interface is of the order of $1-2$ Å.

An in situ electrochemical technique that has been established recently is the quartz microbalance: this instrument can measure small changes in the mass of a metallic electrode that is attached to a quartz oscillator.[14] In this way the electrosorption valency[15] can be calculated directly from the amount of charge from the voltammogram, and the mass obtained from the microbalance. Proper interpretation of the results of this instrument requires electrode surfaces that have large molecularly smooth regions.

A further method that has yielded very interesting information about the structure and interactions in the diffuse part of the double layer is the direct measurement of forces between colloidal particles.[16] Here the forces between two mica plates are measured in the presence of different solutions. Quite interestingly the forces give raise to oscillations of a period similar to the dimensions of the molecules enclosed between the mica plates.

Amongst the optical techniques are also the more traditional methods such as ellipsometry, electroreflectance, and particularly surface plasmons,[17] where experimental and theoretical advances[18] have made it possible to offer a picture of the surface electronic states of the metal in some selected cases, such as the silver (111) phase. We should mention here the measurement of image potential induced surface states by electroreflectance spectroscopy.[19] In this case, besides the normal surface states which arise from the termination of the crystal lattice, there are discrete states due to the existence of an image potential for charges near the conducting interface.

And last, but certainly not least, there is a very extensive literature on the differential capacitance of solutions near either solid (polycrystalline or single crystal) or liquid (mercury) electrodes which we will not try to cover.[20] We should mention the recent work on the influence of the crystallographic orientation of silver on the potential of zero charge of the electrodes, in which a detailed mapping of the influence of the crystal face on the differential capacitance of the inner layer is made.[21]

The complexity of the system described by the experimental methods defies any simple theoretical interpretation. Yet these are needed for the understanding of what is actually going on at the charged interface. It is clear that there are two kinds of forces in these systems: the long-ranged Coulomb forces and the short-ranged forces that are at the origin of the chemical bonds and are also responsible for the repulsion between atomic cores. There are important quantum effects at the interface due essentially to the quantum nature of the electrons in a metal. For this reason we have organized the theoretical discussion of the chapter starting with very simple model systems about which a lot is known, and to systems which are much more realistic, but difficult to handle theoretically. The emphasis of the theoretical treatment will be on the structure functions, or distribution functions $\rho_i(1)$, $\rho_{ij}(1, 2) \ldots$, which give the probability of finding an ion(s) or solvent molecule(s) at specified position(s) near the interface. The properties of the interface can be calculated from these distribution functions.

One of the very interesting theoretical development of recent years has been the exactly solvable model developed by Jancovici, Cornu and co-workers.[22] This is a two-dimensional model at a particular value of the reduced temperature, and is particularly useful to elucidate the subtle properties of the long-ranged Coulomb forces. For the nonprimitive model with solvent molecules there is a one-dimensional exactly solvable model.[23] Exactly solvable models serve as benchmarks for approximate theories and to test exact and general sum rules.

In this chapter we will not discuss the solvent structure, because this is a subject under development. We will restrict ourselves to models in which the solvent is a continuum of dielectric constant ε. The focus of this article is the structure of the inner layer of the interphase, the layer of ions that is directly adsorbed onto the metal surface. We discuss the sticky site model (SSM), in which the adsorption sites are sticky points at the interface. This model requires the distribution functions of the undecorated, smooth surface as an input. The theory of these distribution functions is reviewed in Section I. In Section II we review exact sum rules for the interface. In Section III we discuss the structured interface in the SSM.

I. THEORIES FOR THE SINGLET AND PAIR DISTRIBUTION FUNCTIONS

In recent years there has been significant progress in the statistical mechanics of inhomogeneous charged systems such as the metal electrolyte interface. The real interface is much too complex to be described by a tractable model, so that simple models that focus on some of the more relevant aspects of these systems are used. Most of the effort in the past has been directed to understanding models in which the metal side is an ideally smooth, charged surface, with or without image forces. The solvent is either a continuum of dielectric constant ε, or hard spheres with embedded point dipoles. It is possible to study models with realistic solvents near charged or neutral surfaces in the absence of electrolytes,[24] and less realistic models such as the models with quadrupoles,[25] or with sticky octupoles,[26] in the presence of electrolytes, but for the time being extensive calculations have not been made. The primitive model of charged hard spheres near a charged hard wall, all embedded in a continuum dielectric is by far the most studied, and to date, the best understood of the three-dimensional models of the electric double layer.

There has been considerable progress in the theory of the primitive model of the electric double layer. The theories are constructed as direct improvements on the original Gouy–Chapman theory, such as the Modified Poisson Boltzmann (MPB)[27] theories, the theories based on functional density expansions,[28] and theories derived from integral equations. The MPB theories have been adequately reviewed in another review in this series.[29] The integral equation theories can be classified into three groups: those derived from the Ornstein Zernike (OZ)[30] equation and those derived from the Born–Green–Yvon (BGY),[31] Wertheim–Lovett–Mou–Buff (WLMB),[32] and Kirkwood[33] equations.

As it happens in other areas of condensed matter physics, the forces between molecules is not known, but rather the properties that are measured are interpreted in terms of these forces. Since the equations are only approximate, it is impossible to estimate from the comparison of the theory to experiment the accuracy of the theories. For that reason computer experiments play an important role, since here one knows the input intermolecular potential, and the properties and structure can be computed for systems consisting of a few hundred to a few thousand particles. There are two techniques used to perform these simulations: molecular dynamics (MD), in which the equations of motion of the molecules are solved simultaneously, and the Monte Carlo method (MC), in which an equilibrium ensemble is generated from a random walk algorithm (a Markov Chain). For systems of hard charged spheres the MD technique cannot be used because the forces are both singular and very long ranged, and this produces insurmountable technical problems. The MC techniques are more amenable to simulating the primitive model of the electric double layer,[34] although because of the long range of the Coulomb forces the relatively small size of the system poses problem.[35]

Models in which the solvent is represented by dipolar hard spheres or even higher multipolar solvents have not been simulated in the neighborhood of charged walls. The full solution of the HNC1 equation has not been achieved for the planar wall, only in the the Mean Spherical Approximation (MSA) and in the Generalized Mean Spherical Approximation (GMSA) has been solved.

A. Basic Definitions: The Gouy Chapman Theory

We have a mixture of ions of density ρ_i, charge $e_i = ez_i$, where e is the elementary charge, z_i is the electrovalence, the ion diameter is σ_i, the number density profile of i at a distance z is $\rho_i(z) = \rho_i(1)$ from the electrode, which is always assumed to be flat and perfectly smooth (see Fig. 1).

The singlet distribution function is

$$g_i(z) = h_i(z) + 1,$$
$$= \rho_i(z)/\rho_i, \tag{1}$$

where ρ_i is the bulk density of species i. The charge density $q(z)$ is given by

$$q(z) = \sum_i^m e_i \rho_i(z), \tag{2}$$

where m is the number of ionic species. The electrostatic potential $\phi(z)$ is

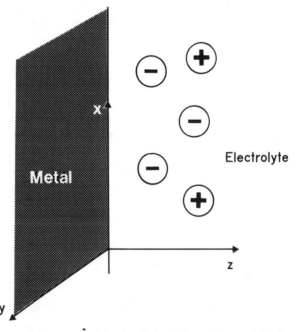

Figure 1. The primitive model of the electric double layer. The metal side is represented by a smooth hard wall. The ions are charged hard spheres. The solvent is a dielectric continuum.

obtained by integration of Poisson's equation

$$\nabla^2\phi(z) = \frac{d^2\phi(z)}{dz^2} = -4\pi q(z)/\varepsilon; \qquad (3)$$

we have

$$\phi(z) = 4\pi \int_z^\infty dt(t-z)q(t)/\varepsilon. \qquad (4)$$

The total potential drop $\Delta\phi$ is obtained from Eq. (4) by either letting $z = 0$ or $z \to -\infty$, depending on the reference potential of the model. In general, the latter choice is adopted. An important quantity is the differentical capacitance C_d which defined by

$$C_d = dq_s/d\Delta\phi \qquad (5)$$

where q_s is the surface charge on the electrode. This quantity is seldom measured directly, it is inferred from either surface tension measurements, or frequency-dependent AC measurements of the capacitance.

The surface charge satisfies the electroneutrality condition

$$q_s = -\int_0^\infty dz\, q(z),$$
$$= E_z(0)\varepsilon/4\pi, \qquad (6)$$

where $E_z(0)$ is the external or applied field.

For low applied field, the Gouy–Chapman–Stern[36,37] theory (or modified Gouy–Chapman Theory, MGC) is accurate for the primitive model in comparison to computer experiments.

Consider the Poisson equation (3). If we approximate the density of the ions by the Boltzmann distribution formula,

$$\rho_i(1) = \rho_i \exp[-\beta e_i \phi(1)], \qquad (7)$$

replacing into Eq. (3), we obtain the Poisson Boltzmann equation

$$\nabla^2\phi(1) = -4\pi/\varepsilon \sum_i e_i \rho_i \exp[-\beta e_i \phi(1)]. \qquad (8)$$

This equation has been solved analytically for several cases of interest. A first integral can be obtained multiplying $\nabla\phi(1)$ both sides of this equation. For planar electrode this yields

$$E_z^2 = [\nabla\phi(1)]^2 = -8\pi kT/\varepsilon \sum_i \rho_i\{\exp[-\beta e_i\phi(1)] - 1\} \qquad (9)$$

Using the definition of C_d and the electroneutrality relation Eq. (6), we get the formula for the differential capacitance

$$C_d = \sqrt{\frac{2\pi}{\varepsilon kT}} \frac{\sum_i e_i \rho_i \exp[-\beta e_i\phi(0)]}{\sqrt{\sum_i \rho_i\{\exp[-\beta e_i\phi(0)] - 1\}}} \qquad (10)$$

where $\phi(0) = \phi(z)|_{z=0} = \Delta\phi$ is the potential at the origin, which is equivalent to the total polarization potential of the electrode.

At this point it is convenient to make a variable change

$$\chi = \exp[-\beta e\phi(0)], \qquad (11)$$

where $e_i = z_i e$, and z_i is the electrovalence of species i.

We integrate Eq. (9) to get

$$\int_{\chi_0}^{\chi(z)} d\chi \frac{1}{\chi\sqrt{\sum_i \rho_i\chi^{z_i} - \rho_0 A}} = \sqrt{\frac{8\pi e^2}{kT\varepsilon}}(z - z_0), \qquad (12)$$

where, for the general case, χ_0 and A are integration constants.[38,39] Since the electrovalences z_i are always small numbers, the integration of the left-hand side is always possible in terms of elliptic functions.[40] For the $z_1 = -z_2 = 1$ case the radicand of the left-hand side is a perfect square and the integral can be performed explicity.

The result for this case is, for the potential drop $\Delta\phi$,

$$\beta e E_0/\kappa\varepsilon = 2\sinh[\Delta\phi e\beta/2]. \qquad (13)$$

The density profile is given by

$$\rho_i(z) = \rho_i\{[1 + z_i\alpha e^{-\kappa z}]/[1 - z_i\alpha e^{-\kappa z}]\}^2 \qquad (14)$$

with $\alpha = \tanh[\Delta\phi e\beta/4]$. κ is the Debye parameter

$$\kappa^2 = 4\pi/kT\varepsilon \sum_i \rho_i e_i^2. \qquad (15)$$

There are several remarks about the Gouy–Chapman theory. From

Eq. (10) and (14) we see that for any *mixture of ions of equal size* the contact theorem is satisfied

$$kT \sum_i \rho_i(0) = \varepsilon/8\pi \, E_z^2(0) + kT \sum_i \rho_i. \tag{16}$$

Contrary to what has been assumed in the literature, the contact theorem is not satisfied for mixtures of unequal size ions. This can be seen from Eq. (16): In the contact theorem all the ions have to be in contact with the wall simultaneously. Since there is only one distance in Eq. (16), there is no way that the theorem will be satisfied. However, for very high positive or negative fields *when only* one kind of ion is present, the Gouy–Chapman theory will satisfy *asymptotically* the all-important contact theorem.

The density profiles obtained from the Gouy–Chapman theory are monotonous, that is, they show no oscillations. When the contact theorem is satisfied, and the electroneutrality integral is correct, $\rho_i(1)$ is pinned at the origin and has a fixed integral, so that the density profile cannot deviate too much from the correct result. However, when the density is high the profiles will be oscillatory, and we should expect deviations from the GC theory. This is also true for the nonprimitive model in which the solvent is a fluid of finite size molecules.

The Gouy–Chapman theory has been solved for nonequal size[41] ions. For 1–1 electrolytes the comparison to computer simulations shows good agreement, because in the regime of high electrode charge there is only one ion present in the double layer. The situation should be different when we are dealing with a mixture of anions or cations of different sizes.

B. Integral Equations–The Primitive Model

The classic Gouy and Chapman[36] theory is based on the Poisson equation and a closure given by Boltzmann's equation. In this theory the ions in the double layer are point charges with no exclusion volume, all embedded in a continuum dielectric. There have been a large number of papers dealing with ways to improve this equation. We must remark, however, that at least in the regime of low density and high temperature (or large dielectric constant), the GC theory is not really bad in spite of its simplifications because it does satisfy the contact theorem Eq. (16) asymptotically for $E \to \infty$, and of course, it also satisfies the electroneutrality condition Eq. (6). However, since the electrochemist is really interested in systems in which the solvent is not a continuum, the primitive model should not be considered a working model for the interpretation of experimental data, but rather a learning model for the theoretician, because of the availability of computer experiments. Indeed as the density and coupling constant increase, significant deviations from the behavior predicted by the GC theory occur.

All of the theories can be formulated as an (integral) equation for the density profile $\rho_i(1)$, or alternatively as a (differential or integrodifferential) equation for the potential $\phi(1)$.[42] Consider first the one-particle direct correlation function, which will be the central quantity of our discussion,[43,44] from which the integral equations can be deduced:

$$c_i(1) = \ln[\rho_i(1)/\mathscr{Z}_i] + \beta u_i(1) \tag{17}$$

where $c_i(1)$ is the one particle direct correlation function, $\mathscr{Z}_i$ is the fugacity of species i, and $u_i(1)$ is the external potential.

The function $c_i(1)$ is a member of the family of direct correlation functions $c(1, 2 \ldots)$,[45] which is the sum of all irreducible graphs with density factors $\rho_i(1)$ for every field point. (For a detailed discussion of correlation functions see, for example, Hansen and McDonald.[46])

1.   Ornstein–Zernike-Based Approximations

At the interface between an electrode and a fluid, the density of the fluid is a function of the distance of the point to the surface $\rho_i(z)$. The Ornstein–Zernike equation for this system can be obtained from an homogeneous mixture in which there are some large ions of radius $R_w \to \infty$, such that $\rho_w R_w^3 \to 0$. In this limit the planar HAB (Henderson–Abraham–Barker) OZ equation is[47]

$$h_i(1) - c_i^w(1) = \sum_j \rho_j \int d2\, c_{ij}^B(|r_{12}|)h_j(2), \tag{18}$$

where $h_i(1)$ is defined by

$$h_i(1) = [g_i(1) - 1] = [\rho_i(1) - \rho_i]/\rho_i. \tag{19}$$

The function $h_i(1)$ is the density profile function for ion i. The function $c_i^w(1)$ is a much more complicated object, and in general does not admit a simple diagram expansion. To gain some insight about the meaning of this function we use functional series expansion.[48]

Consider the functional power series expansion of $\ln[\rho_i(1)]$ around the uniform density ρ_i

$$\beta u_i(1) + \ln[\rho_i(1)] = \ln[\rho_i] + \sum_j \rho_j \int d2\, c_{ij}^B(|r_{12}|)h_j(2)$$

$$+ 1/n! \sum_{jk\ldots} \rho_j\rho_k \ldots \int d2\,d3 \ldots c_{ijk}^B(1,2,3\ldots)h_j(2)h_k(3)\ldots. \tag{20}$$

The direct correlation functions are defined by the functional derivative

$$c_{ijk}^{B}(1, 2, \ldots) = \delta^{n} c_i(1)/\delta \rho_j(2) \delta \rho_k(3) \ldots \tag{21}$$

The superscript B stands for the bulk functions. If we now introduce the new function $c_i^{w}(1)$, defined by

$$c_i^{w}(1) = -\beta u_i(1) - \ln[g_i(1)] + h_i(1)$$

$$+ 1/n! \sum_{ijk} \rho_j \rho_k \ldots \int d2\, d3 \ldots c_{ijk}^{B}(1, 2, 3 \ldots) h_j(2) h_k(3), \tag{22}$$

where the inhomogeneous potential is of the form

$$u_i(1) = u_i^{sr}(1) + w_i(1) \tag{23}$$

where $u_i^{sr}(1)$ is short ranged and for a hard, smooth, charged electrode. The electrostatic part is

$$w_i(1) = -e_i E_z z_1/2. \tag{24}$$

Combining this definition with the functional expansion (Eq. 20) we get the HNC1 equation for the flat wall electrode,

$$-\beta w_i(1) - \ln[g_i(1)] = \sum_j \rho_j \int d2\, c_{ij}^{B}(|r_{12}|) h_j(2). \tag{25}$$

Equation (25) has a deceivingly simple aspect, but in fact, because of the long range of $w_i(1)$, is not convergent, and therefore not amenable to numerical solution. Using the equation

$$c_{ij}^{B}(|r_{12}|) = -\beta w_{ij}(|r_{12}|) + c_{ij}^{sr}(|r_{12}|), \tag{26}$$

with

$$w_{ij}(|r_{12}|) = e_i e_j/\varepsilon r_{12}, \tag{27}$$

and replacing into Eq. (25) yields

$$\beta \phi_i(1) + \ln[g_i(1)] = \sum_j \rho_j \int d2\, c_{ij}^{sr}(|r_{12}|) h_j(2), \tag{28}$$

where $\phi(1)$ is defined by

$$\phi_i(1) = E_z z_1 + \sum_j e_j \int d2\, \rho_j(2)/\varepsilon r_{12}. \qquad (29)$$

This equation is now completely defined in terms of short-ranged quantities, which is not the case for the first form of the equation (25).[49] One important observation about the HNC1 is that it does not satisfy the contact theorem Eq. (16), but rather[50]

$$kT \sum_i \rho_i(0) = \varepsilon/8\pi\, E_z^2(0) + \rho_0 \partial P/\partial\rho_0, \qquad (30)$$

with

$$\rho_0 = \sum_i \rho_i. \qquad (31)$$

However, the HNC1 satisfies automatically the electroneutrality relations and the Stillinger–Lovett sum rules (see discussion below). The HNC is the most accurate theory for bulk electrolytes. It is the theory that has the closure with the largest number of graphs. One would expect that this fact would remain true in the plane electrode limit. However, because of the inaccuracy of the HNC for uncharged hard sphere fluids the HNC1 does not do well in representing the exclusion volume of the ions, and is not on the whole such a good approximation for the electric double layer.

For high fields and low concentrations the fact that we get the compressibility rather than the pressure is not very important and the HNC1 is still a rasonably good theory, as will be shown below. However, for dense systems this is a rather severe shortcoming. Specifically, when we are dealing with a molecular (dipolar) solvent the density is very large and the dielectric constant ε is of the order of one (instead of 80 in water), which makes the electrostatic term in Eq. (16) small in comparison to the contact density term. The consequence is that the HNC1 will put more counter ions near the electrode than the exclusion of the hard cores will permit. Eventually thermodynamic stability conditions will be violated, such as the Bogoliuov inequality.[51] For very high charges the HNC1 predicts a decreasing potential drop ϕ_0 with increasing applied external field E_z.[52]

The bulk direct correlation function

$$c_{ij}^B(|r_{12}|) \qquad (32)$$

which should be used in solving the HNC1 equation (28) is that obtained from the solution of the bulk HNC equation for the same system. This however yields poor results when compared to computer simulations. Generally better results are obtained if instead of the HNC bulk direct correlation function the corresponding MSA functions are used, the general agreement with computer simulations improves.[53]

There have been several calculations with improved versions of this HNC1 equation. Including the next term in the expansion Eq. (22) amounts to including the bridge diagram in the bulk HNC1 calculation. This was done by Ballone et al. with good success.[54] The density profile for the $1M$, $1-1$ electrolyte at a surface charge $\sigma^* = q_s\sigma^2/e = 0.7$, which will be the test case

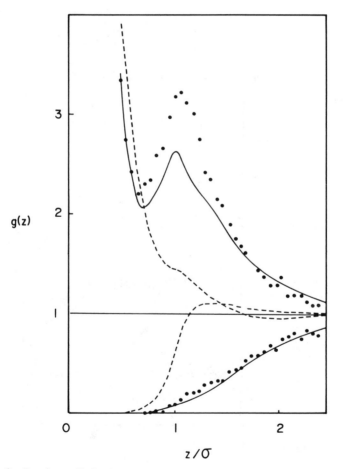

Figure 2. Density profile for the test case with $\sigma^* = 0.7$ (ref. 34, $\sigma^* = \sigma^2 E/e$). The line is the result of Ballone et al.[54]

used for comparisons. This is the highest surface density simulated, and shows charge oscillations due to the hard core of the electrolyte. In this calculation the bridge diagrams were computed directly from their definition.

$$c_i^w = -\beta u_i(1) - \ln[g_i(1)] + h_i(1)$$

$$+ 1/2 \sum_{ijk} \rho_j \rho_k \int d2\, d3\, h_{ij}^B(12) h_{ik}^B(13) h_{kj}^B(32) h_j(2) h_k(3), \tag{33}$$

where the product of the three bulk pair correlation functions in the second term of the right-hand side is the first term in the density expansion of

$$c_{ijk}^B(1, 2, 3). \tag{34}$$

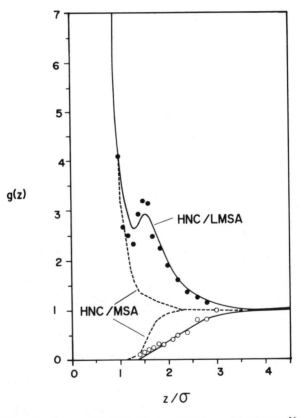

Figure 3.   Same as Fig. 2. Results from Nielaba and Forstmann.[56]

The results of this calculation are shown in Fig. 2. Since there are no adjustable parameters, the agreement is very good. An alternative, less laborious procedure was suggested by Rosenfeld and Blum,[55] but actual calculations were not performed.

Another way of improving the HNC1 approximation was introduced by Forstmann and co-workers.[56] In their method the HNC1 equation is used as described above, but instead of taking the bulk direct correlation function, as prescribed by Eq. (28), a local density dependent $c^B(r, \bar{\rho})$ is taken. The local density is defined by

$$\bar{\rho}_i(x) = \frac{1}{2\sigma\Delta} \int_{x-\Delta}^{x+\Delta} dz \int_{z-\sigma/2}^{z+\sigma/2} dz \, \rho_i(y), \tag{35}$$

where σ is the diameter of the ion and Δ is an adjustable parameter. The bulk correlation function is then

$$c_{ij}^B(|r_{12}|)|_{\rho_i=\bar{\rho}_i}. \tag{36}$$

For the test case with surface charge $\sigma^* = 0.7$, the results of this method, as shown in Fig. 3, are very good.

2. BGY-Based Approximations

The BGY equation can be derived from the one-particle direct correlation function $c_i(1)$. Consider again Eq. (20): letting the gradient ∇ act on the f bonds of the graphical expansion of $c_i(1)$,[57] we get the BGY equation

$$-kT\vec{\nabla}_1\rho_i(1) = \rho_i(1)\vec{\nabla}_1 u_i(1) + \sum_j \int d2\, \rho_{ij}(1,2)\vec{\nabla}_1 u_{ij}(1,2). \tag{37}$$

Using Eqs. (23) and (26) to eliminate the long-range terms, we obtain Eq. (37) in a different form:

$$
\begin{aligned}
-kT\vec{\nabla}_1\rho_i(1) = {} & \rho_i(1)\vec{\nabla}_1 u_i^{sr}(1) + \rho_i(1)e_i\vec{\nabla}_1\phi_i(1) \\
& + \sum_j \int d2\, \rho_{ij}(12)\vec{\nabla}_1 u_{ij}^{sr}(12) \\
& + \rho_i(1)e_i \sum_j \int d2\, \rho_j(2)e_j h_{ij}(12)\vec{\nabla}_1(1/\varepsilon r_{12}).
\end{aligned} \tag{38}
$$

This equation can be integrated from ∞ to z to yield[42]

$$\ln[g_i(z)] = -e_i[\phi(z) + \psi_i(z)] + J_i(z), \tag{39}$$

which together with the Poisson equation (Eq. 3) forms a closed system of equations that is very convenient for numerical solutions.

This equation is of the same type as the one derived from the HNC1 equation (Eq. 28). The right-hand side term consists of three contributions: the potential $\phi(1)$, which is determined by the single-particle distribution function $\rho_i(1)$, and the terms $\psi_i(z)$ and $J_i(z)$, which are functions of the pair distribution function $h_{ij}(1,2)$. From Eq. (38) we get

$$\psi_j(z) = \int_z^\infty dz_1\, \rho_i(1) \sum_j \int d2\, \rho_j(2) e_j h_{ij}(1,2) \vec{\nabla}_1 (1/\varepsilon r_{12}),$$

$$J_i(z) = \int_z^\infty dz_1 \sum_j \int d2\, \rho_{ij}(12) \vec{\nabla}_1 u_{ij}^{sr}(1,2).$$

$$(40)$$

We remark that in Eq. (38) (and also in Eq. 39), if the fluctuation terms $J_i(z)$ and $\psi_i(z)$ are neglected, we get back the Gouy Chapman theory, which has a known analytical solution.

In the BGY-based theories the pair correlation function $h_{ij}(1,2)$ must be given by some approximation. The interesting feature of the BGY equation is that for no matter which closure, the contact theorem (Eq. 16) is satisfied. However, the electroneutrality conditions (see below) are not satisfied in general. The simplest approximation one could think of which is equivalent to Kirkwoods superposition consists in writing

$$h_{ij}(1,2) = h_{ij}^B(r_{12}),$$

$$(41)$$

where $h_{ij}^B(r_{12})$ are the bulk pair correlation functions, fails to satisfy the important electroneutrality condition (see below)

$$-e_i = \sum_j \int d2\, \rho_j(2) e_j h_{ij}(1,2),$$

$$(42)$$

and gives very poor results when compared with the computer simulations. The approximation[58]

$$h_{ij}(12) = f_i(1) f_j(2) h_{ij}^B(r) \qquad r_{12} > \sigma_{ij},$$

$$\qquad\qquad = -1 \qquad\qquad\qquad r_{12} < \sigma_{ij},$$

$$(43)$$

where

$$\sigma_{ij} = (\sigma_i + \sigma_j)/2$$

$$(44)$$

is constructed so that the functions $f_i(z)$ are required to satisfy the electroneutrality condition (Eq. 42) for the inhomogeneous pair distribution function. This yields an integral equation for those functions. This construct of the inhomogeneous pair correlation function can give negative values of $g_{ij}(1, 2)$. The problem is especially severe for dense systems, beyond $2M$ of salt concentration, or also for the test case of high surface density charge. A simple way to circumvent this problem was suggested by Caccamo et al.[59]

$$h_{ij}(1, 2) = f_i(1)f_j(2)h_{ij}^B(r) + A_{ij}(1, 2) \qquad r_{12} > \sigma_{ij},$$

$$= -1 \qquad r_{12} < \sigma_{ij}, \qquad (45)$$

where

$$A_{ij}(1, 2) = Ah_{ij}^{PY}(r_{12}) \qquad z_1, z_2 < 2\sigma_{ij}. \qquad (46)$$

The parameter A is adjusted to eliminate negative correlations from $h_{ij}(1, 2)$. The results are, however, not very sensitive to the exact value of A. The comparison to the Monte Carlo simulations is again good; this method yields for the test case with $\sigma^* = E_z/e\sigma^2 = 0.7$ the density oscillations in the profile of the counterions. Figure 4 shows the comparison to the computer simulations of Torrie and Valleau.[34]

The inhomogeneous pair correlation function $h_{ij}(1, 2)$ can be obtained from the inhomogeneous OZ equation

$$h_{ij}(1, 2) - c_{ij}(1, 2) = \sum_k \int d3\, h_{ik}(1, 3)\rho_k(3)c_{kj}(3, 2) \qquad (47)$$

and a suitable closure for the direct correlation function.

$$c_{ij}(1, 2) = -\beta u_{ij}(1, 2) \qquad r_{12} > \sigma_{ij} \quad \text{(MSA2 approximation)}, \qquad (48)$$

$$c_{ij}(1, 2) = -\beta u_{ij}(1, 2) + h_{ij}(1, 2) - \ln[g_{ij}(1, 2)] \qquad r_{12} > \sigma_{ij}$$

$$\text{(HNC2 approximation)} \qquad (49)$$

The MSA2 approximation cannot be integrated explicitly, as is the case of the homogeneous MSA. When the ions are approximated by charged points, then for some specific form of the density profiles $\rho_i(1)$ the OZ equation can be integrated, and series solutions have been given (Blum et al.[42] and Carnie and Chan[60]). The numerical solution of the MSA2 and HNC2 has been extensively studied by Plischke and Henderson.[61] Approximate solutions of this kind will satisfy the contact theorem (Eq. 16) as well as the electroneutrality and dipole sum rules.[62]

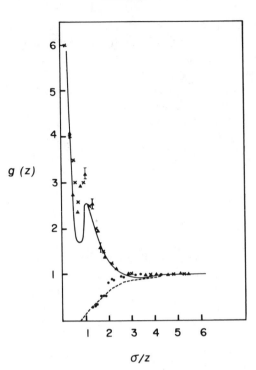

Figure 4. Same as Fig. 2. Results from Caccamo, et al.[59]

The BGY-HNC2 equation has been solved numerically by Nieminen, Ashcroft and collaborators,[63] however for systems with neutral molecules.

3. WLMB-Based Equations

Yet another integral equation is derived from the one-particle direct correlation function $c_i(1)$ (Eq. 22) by introducing relative coordinates for all fields in the diagram representation and taking the derivatives with respect to those coordinates.[57,64] This yields an exact hierarchy of equations that is related to the BGY hierarchy. The first member of the Wertheim–Lovett–Mou–Buff (WLMB) equation is

$$\vec{\nabla}_1 \rho_i(1) + \beta \rho_i(1) \vec{\nabla}_1 u_i(1) = \rho_i(1) \sum_j \int d2 \, c_{ij}(1,2) \vec{\nabla}_2 \rho_j(2). \tag{50}$$

This equation contains long-range, divergent terms. Introducing the local

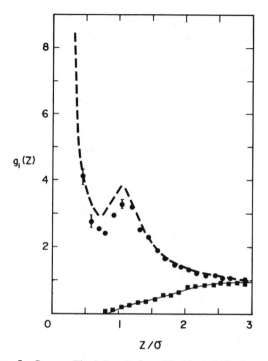

Figure 5. Same as Fig. 2. Results from Plischke and Henderson.[66]

potential $\phi(1)$ (Eq. 29), we have,[65,66]

$$\vec{\nabla}_1 \ln[\rho_i(1)] + \beta\vec{\nabla}_1\phi(1) + \beta\vec{\nabla}_1 u_i^{sr}(1) = \sum_j \int d2\, c_{ij}^{sr}(1,2)\vec{\nabla}_2\rho_j(2). \qquad (51)$$

This equation has been studied by Plischke and Henderson[66] in detail. As can be seen in Fig. 5, it yields very good results for the test case of $\sigma^* = 0.7$. The calculations were performed solving both the HNC2 closure for the inhomogeneous pair correlation function, and also the MSA2 closure in a few cases.

A simplified version of the WLMB equation that produces reasonably good results was studied by Colmenares and Olivares.[67]

4. Kirkwoods Equations

An interesting approach has been suggested by Kjellander and Marcelja,[68] based on the observation that for the HNC approximation the chemical potential can be obtained explicitly as a function of the pair potential $h_{ij}(r_{12})$ for an homogeneous fluid. Then, within the HNC the function $c_i(1)$ can be explicitly evaluated.

The central idea is to slice the three-dimensional space into two-dimensional layers that are homogeneous. The three-dimensional OZ equation can be mapped into a coupled set of N two-dimensional OZ equations for a mixture of N components; each component is an ion in a different layer. The particles interact with a species-dependent interaction pair potential. In the limit of an infinite number of layers this procedure yields the correct inhomogeneous OZ equation. The chemical potential $\mu_i(\alpha)$ of the ith ion in the αth layer is given by Kirkwoods equation:

$$\mu_i(\alpha) = kT \ln \rho_i(\alpha) + kT \ln \Lambda_0 / \Delta z + V_i(\alpha)$$

$$+ \sum_{\beta, j} \rho_i(\beta) \int_0^1 d\lambda \int d\vec{R} g_{ij}(R, \alpha\beta; \lambda) \frac{\partial g_{ij}(R, \alpha\beta; \lambda)}{\partial \lambda}, \tag{52}$$

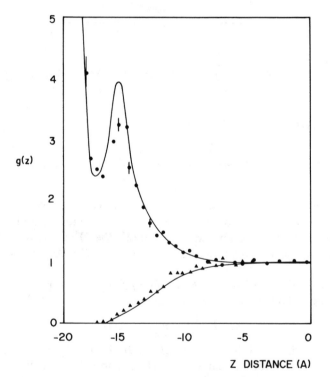

Z DISTANCE (A)

Figure 6. Same as Fig. 2. Results from Kjellander and Marcelja.[68]

where λ is the coupling parameter, Δz is the thickness of the layer, Λ_0 is the ideal gas fugacity, $V_i(\alpha)$ is the interaction between a particle in layer α and the walk, and R is the two-dimensional distance. In the HNC closure

$$c_{ij}(R, \alpha\beta) = h_{ij}(R, \alpha\beta) - u_{ij}(R, \alpha\beta)/kT - \ln[g_{ij}(R, \alpha\beta)]. \tag{53}$$

Kirkwoods equation can be integrated to yield

$$\rho_i(\alpha) = \Delta z/\Lambda_0 \exp\left\{\beta\mu_i(\alpha) + \sum_{\beta,j} \rho_i(\beta) \int d\vec{R}\right.$$

$$\times \left[1/2h_{ij}^2(R, \alpha\beta) - c_{ij}(R, \alpha\beta) - u_{ij}(R, \alpha\beta)/kT \right]$$

$$\left. - 1/2 \ln[g_{ij}(R, \alpha\beta)/\exp[u_{ij}(R, \alpha\beta)/kT]]_{R=0} - \Phi_i(\alpha), \right. \tag{5.4}$$

where $\Phi_i(\alpha)$ is the average potential for layer α,

$$\Phi_i(\alpha) = 2\pi e^2/\varepsilon \sum_\beta \rho_i(\beta)|z_\alpha - z_\beta|. \tag{55}$$

The results for the same case of $\sigma^* = 0.7$ are shown in Fig. 6. The agreement is also excellent.

II. SUM RULES FOR THE CHARGED INTERFACE

The sum rules for the charged interfaces can be classified in two categories:

1. The dynamic sum rules, which are derived from balance of forces considerations.

2. The screening sum rules, which are specific to Coulomb forces. Because of the very long range of the electrostatic forces, the stability of the system requires that all charges surround themselves with a neutralizing cloud.

A. Dynamic Sum Rules

Consider the Born–Green–Yvon (BGY) equation[33,69,70]

$$-kT\vec{\nabla}_1\rho_i(1) = \rho_i(1)\vec{\nabla}_1 u_i(1) + \sum_j \int d2\, \rho_{ij}(1,2)\vec{\nabla}_1 u_{ij}(1,2), \tag{56}$$

where k is Boltzmann's constant, T is the absolute termperature, $\rho_i(1)$ is the density of species i at position $r_1 \equiv x_1, y_1, z_1$. If species i is nonspherically symmetric, the integration should also include the rotational coordinates, the

Euler angles $\Omega_1 \equiv \alpha_1 \beta_1 \gamma_1 u_i(1)$ is the single-body potential acting on i. The pair potential is represented by $u_{ij}(1,2)$. Similarly $\rho_{ij}(1,2)$ is the pair density function, which can be written in the form

$$\rho_{ij}(1,2) = \rho_i(2)\rho_j(2)g_{ij}(1,2), \qquad (57)$$

where $g_{ij}(1,2)$ is the pair distribution function.

For charged particles it is convenient to separate the electrostatic from the nonelectrostatic contributions. For the singlet potential we have

$$u_i(1) = e_i\phi(1) + u_i^{sr}(1), \qquad (58)$$

where $\phi(1)$ is the local potential at $\mathbf{r}_1$, e_i is the electric charge of species i and $u_i^{sr}(1)$ is the sum of all nonelectrostatic forces, which includes the hard repulsion from the wall. Similarly, the two-body contribution is

$$u_{ij}(1,2) = e_i e_j/\varepsilon r_{12} + u_{ij}^{sr}(1,2), \qquad (59)$$

where $u_{ij}^{sr}(1,2)$ is the nonelectrostatic interaction of the pair ij, and the first term is the electric potential between the charges of i and j. Furthermore, we use the function

$$h_{ij}(1,2) = g_{ij}(1,2) - 1 \qquad (60)$$

and the potential

$$\Psi(1) = \phi(1) + \sum_j e_j \int d2\, \rho_j(1)/\varepsilon r_{12}, \qquad (61)$$

where $r_{12} = |\mathbf{r}_1 - \mathbf{r}_2|$. The BGY equation is then transformed to

$$-kT\vec{\nabla}_1(1) = \rho_i(1)\vec{\nabla}_1 u_i^{sr}(1) + \rho_i(1)e_i\vec{\nabla}_1\Psi_i(1)$$

$$+ \sum_j \int d2\, \rho_{ij}(1,2)\vec{\nabla}_1 u_{ij}^{sr}(1,2)$$

$$+ \rho_i(1)e_i \sum_j \int d2\, \rho_j(2)e_j h_{ij}(1,2)\vec{\nabla}(1/\varepsilon r_{12}). \qquad (62)$$

We consider a system which is limited by an arbitrarily rough planar, but charged surface. The precise mathematical requirement is that there is a prism with an arbitrarily large cross-sectional area S and height L (the volume

$V = SL$), such that the force through the walls parallel to z is of $O(S^{1-\delta})$, where $\delta > 0$ as $S \to \infty$.

We integrate now over the whole prism of volume V. This is a generalization of the case of a decorated surface recently discussed in the literature,[71] and is of interest in the discussion of the electrochemistry of rough surfaces.

We integrate the BGY equation in the volume of a prism of the same section S but smaller height $L_1 < L$. Summing over all species i, we get

$$-kT \sum_i \int d1\, \vec{\nabla}_1 \rho_i(1) = \sum_i \int d1 [\rho_i(1)\vec{\nabla}_1 u_i^{sr}(1) + \rho_i(1)e_i\vec{\nabla}_1 \psi_i(1)]$$

$$+ \sum_i \int d1 \sum_j \int d2 [\rho_{ij}(1,2)\vec{\nabla}_1 u_{ij}^{sr}(1,2)$$

$$+ \rho_i(1)\rho_j(2)h_{ij}(1,2)\vec{\nabla}_1 (e_i e_j/\varepsilon r_{12})]. \tag{63}$$

Since in our system $\vec{\nabla}_1 = \partial/\partial z_1$, the integral of the left-hand side term can be easily performed:

$$\int d\mathbf{r}_1 \vec{\nabla}_1 \rho_i(1) = \int_S dx_1\, dy_1 \int_{z_s(x_1,y_1)}^{L_1} dz_1\, \partial\rho_i(1)/\partial z_1$$

$$= S \left[\rho_i(L_1) - 1/S \int dx_1\, dy_1\, \rho_i(x_1,y_1,z_s) \right] \tag{64}$$

$$\int d\mathbf{r}_1 \vec{\nabla}_1 \rho_i(1) = S[\rho_i(L_1) - \bar{\rho}_i(0)],$$

where we have defined the average contact value as

$$\bar{\rho}_i(0) = 1/S \int dx_1\, dy_1\, \rho_i(x_1, y_1, z_s). \tag{65}$$

To integrate the second term in the right-hand side we use Poisson's equation

$$\nabla_1^2 \psi(1) = -4\pi/\varepsilon \sum_i e_i\rho_i(1) = -\vec{\nabla}_1 \cdot \mathbf{E}(1), \tag{66}$$

where we have used the electric field

$$\mathbf{E}(1) = -\vec{\nabla}_1 \psi(1). \tag{67}$$

Substitution into the second term of the right-hand side leads to

$$\int d\mathbf{r}_1 [\sum e_i \rho_i(1)][E_z(1)] = -\varepsilon/4\pi \int d\mathbf{r}_1 \nabla_1^2 \psi(1) E_z(1)$$

$$= \varepsilon/4\pi \int d\mathbf{r}_1 [\vec{\nabla}_1 \cdot \mathbf{E}(1)] E_z(1), \qquad (68)$$

where $\mathbf{E}(1)$ is the electric field at position $\mathbf{r}_1$, and $E_z(1)$ is the z component. Integrating by parts, we get

$$\int d\mathbf{r}_1 \left[\sum e_i \rho_i(1) \right][E_z(1)] = -\varepsilon/8\pi \int_S dx_1 \, dy_1 \int_{z_s(x_1,y_1)}^{L_1} dz_1 \, \partial E_z^2(1)/\partial z$$

$$+ \varepsilon/4\pi \int dx_1 \, dy_1 \, dz_1 [E_z(1)\partial E_x(1)/\partial x_1 + E_z(1)\partial E_y(1)/\partial y_1]. \qquad (69)$$

The second term of the right-hand side is zero: for a periodic interface in the x and y directions, if we take S to be the surface of a unit cell, then the terms like $\partial E_x(1)/\partial x_1$ will be equal but with opposite sign for neighboring cells. For the general random interface we conjecture that this term is finite: then in the limit $S \to \infty$ the contribution vanishes. We have

$$1/S \int d\mathbf{r}_1 \left[\sum e_i \rho_i(1) \right][E_z(1)] = -\varepsilon/8\pi \langle E_z^2 \rangle_S, \qquad (70)$$

where the average square field in the z direction is

$$\langle E_z^2 \rangle_S = 1/S \int_S dx_1 \, dy_1 \, E_z^2(x_1, y_1, z_s). \qquad (71)$$

The first term yields

$$1/S \int d1 \, \rho_i(1)\vec{\nabla}_1 u_i^{sr}(1) = \langle \rho_i(1)\partial u_i^{sr}(1)/\partial z_1 \rangle_S, \qquad (72)$$

where

$$\langle \rho_i(1)\partial u_i^{sr}(1)/\partial z_1 \rangle_S = 1/S \int_S dx_1 \, dy_1 \int_{z_s(x_1,y_1)}^{\infty} dz_1 \, \rho_i(1)\partial u_i^{sr}(1)/\partial z_1. \qquad (73)$$

The last term in the right-hand side of Eq. (63) is the average of the pair forces. We write

$$I_p = \int_S dx_1\,dy_1 \int_S dx_2\,dy_2 \int_{z_s(x_1,y_1)}^{L_1} dz_1 \int_{z_s(x_2,y_2)}^{L} dz_2 \sum_{ij} F_{ij}(1,2)$$

$$= \int_S dx_1\,dy_1 \int_S dx_2\,dy_2 \int_{z_s(x,y)}^{L_1} dz_1 \left[\int_{z_s(x,y)}^{L_1} dz_2 + \int_{L_1}^{L} dz_2 \right] \sum_{ij} F_{ij}(1,2),$$

$$(74)$$

where

$$F_{ij}(1,2) = [\rho_{ij}(1,2)\vec{\nabla}_1 u_{ij}^{sr}(1,2) + \rho_i(1)\rho_j(2)h_{ij}(1,2)\vec{\nabla}_1(e_i e_j/\varepsilon r_{12})]. \qquad (75)$$

Clearly,

$$F_{ij}(1,2) = -F_{ji}(2,1) \qquad (76)$$

from where the first double integral in Eq. (74) vanishes. Since the interactions are short-ranged, the second double integral yields the virial contribution to the bulk pressure P,

$$P = 1/S \int_S dx_1\,dy_1 \int_S dx_2\,dy_2 \int_{z_s(x,y)}^{L_1} dz_1 \int_{L_1}^{L} dz_2 \sum_{ij} F_{ij}(1,2)$$

$$= 2\pi/3 \sum_{ij} \rho_i \rho_j \int_0^\infty dr\, r^3 [e_i e_j h_{ij}(r)/\varepsilon r^2 + g_{ij}(r)\partial u_{ij}(r)/\partial r]. \qquad (77)$$

Putting it all together yields the general contact theorem for a surface that is planar on the average, but not necessarily smooth:

$$kT \sum_i \bar{\rho}_i(0) = P + \varepsilon/8\pi\langle E_z^2\rangle_S - \sum_i \langle \rho_i(1)\partial u_i^{sr}(1)/\partial z_1 \rangle. \qquad (78)$$

This theorem is a generalization of the previously derived contact theorems to the realistic case of nonsmooth electrode surfaces. It contains the previous results as particular cases. If the interface is a smooth hard wall, the surface averages become the surface values of the parameters and we get

$$kT \sum_i \rho_i(0) = P + \varepsilon/8\pi E_z^2(0) - \sum_i \langle \rho_i(1)\partial u_i^{sr}(1)/\partial z_1 \rangle, \qquad (79)$$

where the last term is now

$$\langle \rho_i(1)\partial u_i^{sr}(1)/\partial z_1 \rangle = \int_0^\infty dz_1\, \rho_i(1)\partial u_i^{sr}(1)/\partial z_1, \qquad (80)$$

when $u_i^{sr}(1)$ is zero and we recover the previously obtained results

$$kT \sum_i \rho_i(0) = P + \varepsilon/8\pi E_z^2(0) \tag{81}$$

for the primitive model with a continuum solvent of dielectric constant ε, and

$$kT\left[\sum_i \rho_i(0) + \sum_n \rho_n(0)\right] = P + 1/8\pi E_z^2(0) \tag{82}$$

for the molecular solvent case. Here the second sum in the left-hand side is over all the neutral molecules in the system, while the first is over all the ions present.

When the surface has an array of sticky adsorption sites, such as in the case of the SSM model discussed below, the adsorption potential has the form

$$\exp[-\beta u_a(\mathbf{r})] = 1 + \lambda_a(\mathbf{R})\delta(z), \tag{83}$$

with

$$\lambda_a(\mathbf{R}) = \sum_{n_1 n_2} \lambda_a \delta(\mathbf{R} - n_1 \mathbf{a}_1 - n_2 \mathbf{a}_2). \tag{84}$$

Here $\mathbf{R} = x, y$ is the position at the electrode surface, and z the distance to the contact plane, which is at a distance $\sigma/2$ from the electrode. In Eq. (84), n_1, n_2 are natural numbers, and $\mathbf{a}_1, \mathbf{a}_2$ are lattice vectors of the adsorption sites on the surface. The number λ_a represents the fugacity of an adsorbed atom of species a. Define now the regular part of the density function

$$\rho_i(1) = y_i(1)\exp[-\beta u_i^{sr}(1)]. \tag{85}$$

Replacing into the general contact theorem (Eq. 78) yields

$$kT \sum_i \bar{\rho}_i(0) = P + \varepsilon/8\pi\langle E_z^2\rangle_S + kT\lambda_a \sum_i \langle \partial y_i(1)/\partial z_1\rangle. \tag{86}$$

This theorem has been verified recently by Cornu, for the exactly solved model of a one-component plasma in two dimensions.[72]

B. The Screening Sum Rules

In systems that are electrically neutral, any fixed arrangement of charges is screened by the mobile charges of the system. In homogeneous bulk phase this is an intuitively natural fact, because if the long ranged Coulomb forces are not screened, the partition function would not exist (it would diverge), and matter would not be stable. This is expressed by the fact that the charge distribution around a given charge e_i is of equal value but opposite sign. In the homogeneous bulk phase

$$-e_i = \sum_j \int d2 \, \rho_j(2) e_j h_{ij}(1,2) \tag{87}$$

Rotational invariance in bulk fluids requires that not only charges but also multipole of arbitrary order should be screened by the mobile charges of the media.[73] This fact is much less intuitive in the neighborhood of charged objects, in particular in the neighborhood of a charged electrode. However, the theorems hold and in classical mechanics, at least, perfect screening of all multipoles occurs, in the homogeneous or inhomogeneous systems.

These conclusions are supported by the results of the exactly solved Jancovici model.[74] However, perfect screening of all multipoles does not occur in quantum systems or in systems out of equilibrium.[75]

As a consequence of the screening the second moment of the pair distribution function must be normalized. This is the Stillinger–Lovett moment relation.[76] As was shown by Outhwaite,[77] it can be written in the form of a normalization condition for the electrostatic potential

$$\psi_i(\mathbf{r}) = 1/\varepsilon \left[e_i/r + \sum_j \rho_j e_j \int dr_1 \, h_{ij}(r_1)/(|\mathbf{r} - \mathbf{r}_1|) \right]. \tag{88}$$

When have then that the second moment relation is

$$1/kT \left[\sum_j \rho_j e_j \int d\mathbf{r}_1 \psi_j(1) \right] = 1. \tag{89}$$

Carnie and Chan[78] have shown that this normalization condition is also valid for the inhomogeneous systems of charged particles. Consider the inhomogeneous Ornstein–Zernike (OZ) equation

$$h_{ij}(1,2) - c_{ij}(1,2) = \sum_k \int d3 \, h_{ik}(1,3) \rho_k(3) c_{kj}(3,2), \tag{90}$$

where $c_{ij}(1,2)$ is the direct correlation function and the singlet density function $\rho_i(1)$ satisfies the electroneutrality condition

$$\sum_j e_j \int d\mathbf{r}_1 \, \rho_i(1) = -q_s S, \tag{91}$$

where q_s is the surface charge density and S is the area of the interface. The pair distribution function satisfies the sum rule

$$-e_i = \sum_j \int d2\, \rho_j(2) e_j h_{ij}(1,2). \tag{92}$$

From the diagram expansion we write the direct correlation function as

$$c_{ij}(1,2) = c_{ij}^0(1,2) - \beta u_{ij}^{el}(1,2), \tag{93}$$

where

$$u_{ij}^{el}(1,2) = e_i e_j / \varepsilon r_{12} \tag{94}$$

is the electrostatic part of the interaction and $c_{ij}^0(1,2)$ is the short-range part of the direct correlation function. In the dense media

$$\phi_i(1,2) e_j = -u_{ij}^{el}(r_{12}) + \sum_k \int dr_3 u_{ik}^{el}(r_{13}) \rho_k(r_3) h_{kj}(3,2), \tag{95}$$

and from the OZ equation (Eq. 90) we get

$$h_{ij}(1,2) = -\beta \phi_i(1,2) e_i - c_{ij}^0(1,2) + \sum_k \int d3\, c_{ik}^0(1,3) \rho_k(3) h_{kj}(3,2). \tag{96}$$

Multiplying this equation by $e_j \rho_j(2)$, integrating over $\mathbf{r}_2$ and summing over j yields, after use of the electroneutrality condition (Eq. 92),

$$1/kT \sum_i e_i \int d\mathbf{r}_1 \rho_i(1) \phi_i(1,2) = 1, \tag{97}$$

which is the generalization of the Outhwaite formula of the Stillinger–Lovett sum rule for inhomogeneous charged systems.

C. Other Sum Rules

For flat hard electrode surfaces there are number of other sum rules. A complete review of these rules was recently made by Martin.[79] We will just mention a few of those more relevant to the calculation of density profiles in the electric double layer.[80]

$$kT\partial \ln \rho_i(1)/\partial E_0 = \sum_j \int d2\, \rho_j(2) e_j h_{ij}(1,2)(z_2 - z_1), \tag{98}$$

where E_0 is the bare field at the electrode surface. The differential capacity, which is defined by

$$C_d = \partial q_s / \partial \phi_i(1), \qquad (99)$$

where q_s is the surface charge, $q_s = E_0 \varepsilon / 4\pi$, and $\phi(1)$ is the potential drop, satisfies the sum rule

$$1/C_d = 8\pi^2/\varepsilon^2 S \sum_{ij} \int d1\, d2\, \rho_j(2)\rho_i(1)e_j e_i h_{ij}(1,2)(z_2 - z_1)^2. \qquad (100)$$

The surface tension γ obeys relations that can be given in terms of the direct correlation function[81,82]

$$\gamma = (\pi k T/2) \sum_{ij} \int dz_1\, dz_2\, \partial \rho_i(1)/\partial z_1\, \partial \rho_j(1)/\partial z_2 \int dr_{12}\, r_{12}^3 c_{ij}(1,2). \qquad (101)$$

Using the OZ equation we get the form that contains the pair correlation function

$$\gamma = (\pi/2kT) \sum_{ij} \int dz_1\, dz_2\, \rho_i(1)\partial u_i(1)/\partial z_1\, \rho_j(2)\partial u_j(1)/\partial z_2 \int dr_{12} r_{12}^3 h_{ij}(1,2). \qquad (102)$$

These sum rules provide ways of asserting the accuracy of the different approximations used to compute the charge and ion density near charged walls.

III. THE STICKY SITE MODEL

The theoretical discussion of smooth surfaces is considerably simpler than that of a realistic surface, in which the solid, usually a metal, has a well-defined crystal structure. The reason is that in the case of a smooth surface the problem is one-dimensional, rather than three-dimensional. The analysis of a realistic metal surface potential in contact with an ionic solution is extremely difficult, and requires the use of very large computers. We would like to discuss a simple model of a structured interface which predicts surface phase behavior for the adsorbed layers and which is mathematically tractable. In fact, if the correlation functions of the smooth surface model are known to all orders, then the properties and correlation functions of our model can be computed exactly.

The model[83,84] combines two ideas that were used a very long time ago: Boltzmann's sticky potential[85] and the adsorption site model of Langmuir.[86]

The elegant work of Baxter[87] in which the Percus–Yevick approximation of the sticky hard sphere model is solved and discussed, shows that this model has a particularly simple mathematical solution. In Baxter's work the potential has the form

$$\exp[-\beta u_a(\mathbf{r})] = 1 + \lambda\delta(r - \sigma^-), \tag{103}$$

where $\beta = 1/kT$ is the usual Boltzmann thermal factor, $u(\mathbf{r})$ is the intermolecular potential, λ is the stickiness parameter, $\mathbf{r} = (x, y, z)$ is the relative position of the center of the molecules, and σ is the diameter of the molecules. The right-hand side term represents the probability of two molecules being stuck by the potential $u_a(\mathbf{r})$: this occurs only when the two molecules actually touch, and for this reason we use the Dirac delta function $\delta(r - \sigma^-)$, which is zero when the molecules do not touch and infinity when they do, but the integral is normalized to one. The stickiness is represented by the parameter λ, which, except for a normalization factor, can be considered as the fugacity of the formation of the pair.

The Langmuir adsorption sites can be represented by a collection of sticky sites of the same form as was suggested by Baxter. Only now we do not have a sphere covered uniformly by a layer of glue, but rather a smooth, hard surface with sticky points which represent adsorption sites where actual chemical bonding takes place. For this model Eq. (103) has to be changed to

$$\exp[-\beta u_a(\mathbf{r})] = 1 + \lambda_a(\mathbf{R})\delta(z), \tag{104}$$

with

$$\lambda_a(\mathbf{R}) = \sum_{n_1 n_2} \lambda_a\delta(\mathbf{R} - n_1\mathbf{a}_1 - n_2\mathbf{a}_2). \tag{105}$$

Here $\mathbf{R} = (x, y)$ is the position at the electrode surface and z the distance to the contact plane, which is at a distance $\sigma/2$ from the electrode. In Eq. (105) n_1, n_2 are natural numbers and $\mathbf{a}_1, \mathbf{a}_2$ are lattice vectors of the adsorption sites on the surface. The number λ_a represents the fugacity of an adsorbed atom of species a onto the surface, which has a perfectly ordered array of adsorption sites. While point sites are not a requirement of the model, it makes the mathematical discussion much simpler.

Consider now a fluid consisting of only one kind of particles of diameter σ near a smooth, hard wall with sticky sites. The fluid has N particles and the volume of the system is V. The Hamiltonian of the system is

$$H = H^0 + H^2, \tag{106}$$

where H^0 is the Hamiltonian of the system in the absence of the sticky sites on the hard wall and H^2 is the sticky sites interaction.

$$H^2 = \sum_{i=1}^{N} u_a(\mathbf{r}_i) \tag{107}$$

where $u_a(\mathbf{r}_i)$ is the sticky interaction of Eq. (104).
The canonical partition function of this model is[83]

$$Z = \frac{1}{\prod_a N_a!} \int \exp[-\beta H^0] \prod_{i=1}^{N} \prod_a [1 + \lambda_a(\mathbf{R}_i)\delta(z_i)] \, d\mathbf{r}_i. \tag{108}$$

Expanding the product in Eq. (108) and integrating the Dirac delta functions we get, using the single component notation to avoid heavy and unnecessarily complex equations, with the understanding that in the multicomponent case λ is a vector quantity with components $\lambda_1 \ldots \lambda_a$, and the necessary modifications of $N!$ and the integrations have to be made.

$$Z = Z^0 \sum_{\substack{\text{all sites} \\ \text{on the surface}}} (\lambda^s/s!)\rho_s^0(\mathbf{r}_1 \ldots \mathbf{r}_s) \tag{109}$$

where $\mathbf{r}_i = \mathbf{R}_i$, 0 is the position of the ith adsorbing site on the surface and $\rho_s^0(\mathbf{r}_1 \ldots \mathbf{r}_s)$ is the s-body correlation function of the smooth interface model.

$$\rho_s^0(\mathbf{r}_1 \ldots \mathbf{r}_s) = \frac{1}{(N-s)!Z^0} \int \exp[-\beta H^0] \prod_{i=s+1}^{N} d\mathbf{r}_i, \tag{110}$$

$$= g_s^0(\mathbf{r}_1 \ldots \mathbf{r}_s) \prod_{i=1}^{s} \rho^0(\mathbf{r}_i). \tag{111}$$

Equation (111) defines the s-body correlation function $g_s^0(\mathbf{r}_1 \ldots \mathbf{r}_s)$ while $\rho^0(\mathbf{r}_i)$ is the singlet density of the smooth wall inhomogeneous problem. The smooth wall partition function is

$$Z^0 = \frac{1}{N!} \int \exp[-\beta H^0] \prod_{i=1}^{N} d\mathbf{r}_i. \tag{112}$$

The important observation is that in our sticky sites model (SSM), the excess properties of the interface depend only on the correlation functions of the smooth interface model. In fact, introducing the potentials of mean force $w_s^0(\mathbf{r}_1 \ldots \mathbf{r}_s)$,

$$g_s^0(\mathbf{r}_1 \dots \mathbf{r}_s) = \exp[-\beta w_s^0(\mathbf{r}_1 \dots \mathbf{r}_s)], \tag{113}$$

$$Z/Z^0 = \sum_{\substack{\text{all sites} \\ \text{on the surface}}} \frac{\lambda^s \prod_{i=1}^s \rho^0(\mathbf{r}_i)}{s!} \exp[-\beta \sum w_s^0(\mathbf{r}_1 \dots \mathbf{r}_s)]. \tag{114}$$

The left-hand side of Eq. (114) is closely related to the excess free energy of the sticky site model as compared to the smooth wall problem. In fact, from standard thermodynamics, the excess free energy is

$$\Delta F^s = -1/\beta \ln(Z/Z^0). \tag{115}$$

It is clear that in Eq. (114) the right-hand side is the grand canonical partition function of the two-dimensional lattice gas in which the fugacity is $\mathcal{Z}_a = \lambda \rho_a^0(\mathbf{r}_i) = \lambda \rho_a^0(0)$, where we have used the fact that the single particle density of the smooth wall problem depends only on the distance to the electrode z and not on the position of the adsorbed particle on the interface. The SSM is a model that decouples the structure of the interface from the inhomogeneous charged fluid problem, which in itself is a very difficult one, even for the smooth interface. The excess properties of the lattice surface are formally the same as the two-dimensional lattice gas problems that have been extensively investigated, notably in connection with the Ising model.[88] If the area of the interface is S, and the number of sites is M, the number of sites per unit area is

$$\omega = M/S. \tag{116}$$

Define also the excess free energy per unit area

$$\Delta f^s = \Delta F^s/S \tag{117}$$

and the fraction of occupied sites $\theta_a = N_a/M$ of particles of species a. If we compare Eq. (109) to the grand canonical partition function of a lattice gas on the sites of our model, we see that the number of sites occupied by particles of species a is

$$\langle N_a \rangle = \partial \ln Z_N / \partial \ln \lambda_a, \tag{118}$$

and using Eq. (116), the fraction of occupied sites is

$$\theta_a = (1/M) \partial \ln Z_N / \partial \ln \lambda_a. \tag{119}$$

The excess free energy Δf^s is also the excess pressure due to the presence of the discrete sites structure on the interface.

A. Exact Relations for the SSM

1. Sum Rules for the Fluid Density Functions

The SSM for a charged interface satisfies a number of exact relations. The first one is the analogue to the Gibbs absorption equation: if we define the surface excess in the SSM of species a[83]

$$\Delta \Gamma_a = \int_0^\infty dz [\rho_a(z) - \rho_a^0(z)], \tag{120}$$

where $\rho_a^0(z)$ is the density profile of the smooth wall and $\rho_a(z)$ is that of the SSM, then we must have

$$\Delta \Gamma_a = -\beta \partial \Delta f^s / \partial \ln \lambda_a. \tag{121}$$

Because of the singularity in the f function, the distribution functions also must have a singularity. They must be of the form

$$\rho_a(\mathbf{r}) = [1 + \lambda_a(\mathbf{R})\delta(z)] y_a(\mathbf{r}), \tag{122}$$

where $y_a(\mathbf{r})$ is the regular part of the distribution function. The average number of adsorbed particles is given by the integral of the singular part of the density $\rho_a(\mathbf{r})$

$$\langle N_a \rangle = \int \lambda_a(\mathbf{R})\delta(z) y_a(\mathbf{r}) \, d\mathbf{r}, \tag{123}$$

since this integral is actually a sum over the sticky sites m, we get the relation

$$\theta_a = \lambda_a y_a(\mathbf{r}_m), \tag{124}$$

which is exact. The stickiness parameter λ_a is the ratio between the number of adsorbed and nonadsorbed particles at site $\mathbf{r}_m$ of the lattice. Integrating Eq. (122) over all space yields

$$N = \langle N_a \rangle + \int y_a(\mathbf{r}) \, d\mathbf{r}, \tag{125}$$

which can be rewritten as

$$\langle N_a \rangle = -\int d\mathbf{r}[y_a(\mathbf{r}) - \rho_a]. \tag{126}$$

Because of the analogy of the partition function to the grand canonical partition function, a relation similar to the compressibility relation holds,

$$\langle N_a^2 \rangle - \langle N_a \rangle^2 = \lambda_a \frac{\partial \langle N_a \rangle}{\partial \lambda_a}. \tag{127}$$

2. Exact Results for the Adsorbed Layer

Consider the case of a simple salt dissolved in water, near a metallic electrode. In the SSM there will be three components, the anion, the cation, and the solvent, and the lattice atoms. In the limiting case of the SSM, the sizes of the different species play a crucial role in the possible ordering of the ad layers at the interface. It will be convenient to picture the ions as having a hard sphere core with a diameter σ_a, σ_b, and the solvent as having a hard core with diameter σ_n. The lattice spacing of the metal surface is d, and because it is the most stable surface, we will restrict ourselves to discussing the (Eq. 111) surface of the fcc crystals or the (Eq. 100) face of the hcp crystal, that is, the triangular lattice. In the most general case, all three components can be adsorbed competitively and this situation can give rise to very complex phase diagrams.[89] This most general case can be modeled by the spin $S = 1$ Ising model, and has a very rich phase diagram, involving first- and second-order phase transitions and multicritical points. However, in most electrochemical situations the electrode surface is polarized either positively or negatively, which means that either the cation or the anion is strongly repelled from the surface, and therefore we need to consider the adsorption of either a or b and the solvent n on the electrode. This implies a drastic simplification in the model, because now we can discuss at least the case of commensurate adsorption in terms of models that have been solved analytically, such as the spin $S = 1/2$ Ising model and the hard hexagon gas model. The phase transitions predicted for these models seem to be reasonable in terms of the currently available experimental evidence. Consider

$$Z/Z^0 = \sum_{\substack{\text{all sites} \\ \text{on the surface}}} \frac{\lambda^s \prod_{i=1}^{s} \rho^0(\mathbf{r}_i)}{s!} \exp[-\beta \sum w_s^0(\mathbf{r}_1 \dots \mathbf{r}_s)]. \tag{114}$$

Here the sum has two kind of factors, the fugacity of the absorbed molecules

$$\mathscr{L}_a = \lambda_a \rho_a^0(\mathbf{r}), \tag{130}$$

and another factor with corresponds to the interaction of the adsorbed particles

$$g_s^0(\mathbf{r}_1 \ldots \mathbf{r}_s) = \exp[-\beta \sum w_s^0(\mathbf{r}_1 \ldots \mathbf{r}_s)]. \tag{131}$$

Equation (114) is the grand canonical partition function for a lattice gas with arbitrary interactions. In the most general case, the adsorbate could occupy one or more than one single adsorption site, and any adsorbed molecule could interact with an arbitrary number of neighbors. This problem is however quite untractable, and therefore not very useful. The first approximation that comes to mind is the Kirkwood superposition approximation

$$g_s^0(\mathbf{r}_1 \ldots \mathbf{r}_s) = \prod_{1 < i, j < s} g_2^0(\mathbf{r}_i, \mathbf{r}_j). \tag{132}$$

Because of geometrical considerations, this approximation is probably a good one for molecules with short-range interactions such as hard spheres, and not a very good one for unscreened charged particles. In terms of the effective potentials, we have

$$w_s^0(\mathbf{r}_1 \ldots \mathbf{r}_s) = \sum_{1 < i, j < s} w_2^0(\mathbf{r}_i, \mathbf{r}_j). \tag{133}$$

With this approximation the problem can be mapped onto the lattice gas problem with arbitrary interactions, which is still a difficult problem. If we restrict the interactions to nearest neighbors only, then not only is the problem a tractable one, but there is a rather extensive literature on cases that are of physical interest, for which the phase diagram of the two-dimensional lattice gas is known, and hence an adsorption isotherm can be deduced.

There are two cases of chemisorption of electrochemical interest. In the first one, the charge of the adsorbate is neutralized by the electrons in the metal, and this means that the interactions between neighbors on the surface is attractive. If we are far from the point of zero charge, and the metal is negatively charged, the contact probability of the anions a is zero for all practical purposes and we have only the cation b or the solvent n on the adsorption sites. The problem is reduced to a spin s = 1/2 Ising model with ferromagnetic interactions. In the second case there is no discharge of the ad ions by the metal. The interactions between the ions of the same sign is clearly repulsive, so that the nearest neighbor sites to an occupied site are not going to be occupied.

The possible existence of phase transitions was discussed in the work of Huckaby and Blum.[90] The observation is that the SSM model maps the three-dimensional interface onto a two-dimensional lattice problem. The phase behavior in the interphase is determined by the mapping of the parameters, and exact conditions on the existence of phase transitions can be given.

The simplest electrolyte has three components that can be adsorbed onto the sites of the metal: the anion, the cation, and the solvent. The possible phase behavior of such a system has been discussed in the literature and although there are some exact results on the phase behavior of the adsorbed layer, the study of this system remains an open problem. The simpler case of two-component adsorption is discussed in more detail.

B. The Three-State Adsorption Model

This is the case of a simple salt dissolved in a solvent like water. We call the ions a and b and the solvent n. Since this is a dense system then the sticky sites will never be empty, they will be occupied by either a, b, or n. Now there are various possibilities, because the size of adsorbate may be bigger or smaller that the site–site separation. In the first case the adsorbate will exclude not only its own site but also the neighboring ones, and the problem becomes the hard hexagon problem of Baxter when the interactions of the neighboring sites are ignored altogether. If the adsorbate is smaller than the lattice site separation, it can been shown that the problem can be reduced to the two-component lattice gas. This case was recently studied by Rikvold.[89] In fact we have, by direct comparison of the partition function and the Hamiltonian of the three-state lattice gas

$$-(1/\beta)\ln \Xi = -\phi_{aa} \sum_{\langle i,j \rangle} c_i^a c_j^a - \phi_{ab} \sum_{\langle i,j \rangle} (c_i^a c_j^b + c_i^b c_j^a)$$

$$- \phi_{bb} \sum_{\langle i,j \rangle} c_i^b c_j^b - \mu_a \sum_i c_i^a - \mu_b \sum_i c_i^b, \qquad (134)$$

where we have used Rikvold's notation: the operators c_i^x are 1 when site i is occupied by x, and 0 otherwise. In our case x is either a or b, and the solvent n counts as the empty site. The interactions are

$$\phi_{aa} = w_{aa} - w_{nn} = -(1/\beta)\ln(g_{aa}^0(\mathbf{r}_i, \mathbf{r}_j)/g_{nn}^0(\mathbf{r}_i, \mathbf{r}_j)) \quad i, j \text{ are neighboring sites} \tag{135}$$

and similar relations for ϕ_{ab} and ϕ_{bb}. The last two sums in Eq. (134) correspond to the single occupation of a site by x:

$$\mu_a = (1/\beta)\ln \mathcal{Z}_a/\mathcal{Z}_n = (1/\beta)\ln \left[\frac{\lambda_a \rho_a^0(\mathbf{r}_i)}{\lambda_n \rho_n^0(\mathbf{r}_i)} \right] \tag{136}$$

and a similar expression for μ_b.

The possible arrangements on the triangular lattice of a three-component mixture is quite complicated and has been studied recently by Collins et al.[91] (see also Saito[92]). There are 10 different ordered phases on the surface, when only next nearest neighbors are interacting. Of course the picture is even more complicated for longer-ranged interactions. The geometrical arrangement is either (1 × 1), and there are three phases of this kind which correspond to pure a, pure b, and pure solvent n. These are dense phases and correspond to total coverage ($\theta = 1$) for each of these components. These are limiting cases when there is very strong adsorption of any of the components and attractive interaction between atoms of the same kind. If we consider ions with their charge, then because of the Coulombic repulsion the interaction is repulsive or antiferromagnetic and the dense phases with either pure a or pure b are to be excluded. In the case of the underpotential deposited dense monolayers. This is explained by the fact that the metal cations are not charged at the surface, and therefore do not repel each other.

The remaining seven phases are mapped on the ($\sqrt{3} \times \sqrt{3}$) lattice. The unit cell of this lattice contains three sites of the original triangular lattice, and appears rotated by 30° with respect to it. Each of the three sites can be occupied by either of the components of the mixture, but because of the symmetry (geometrical degeneracy) of the phases, the number of distinct coverings is reduced from 24 to 7. In Table I, which is taken directly from Collins's work we show the coverage, energy, and degeneracy of each of the ordered phases.

There are two cases of chemisorption of electrochemical interest. In the first the charge of the adsorbate is neutralized by the electrons in the metal, and this means that the interactions between neighbors on the surface are attractive. If we are far from the point of zero charge, and the metal is negatively charged, the contact probability of the anions a is zero for all practical purposes and we have only the cation b or the solvent n on the adsorption sites. The problem is then reduced to a spin $s = 1/2$ Ising model with ferromagnetic interactions, which will be discussed in the next section.

In the second case there is no discharge of the ad ions by the metal. The interactions between the ions of the same sign is clearly repulsive, and that of opposite sign is attractive. We have then

$$\phi_{ab} < 0, \quad \text{and} \quad \phi_{aa}, \phi_{bb} > 0. \tag{137}$$

This is definitely the antiferromagnetic case. Let us assume for simplicity that we are dealing with the symmetric case in which

$$\phi_{aa} = \phi_{bb}. \tag{138}$$

L. BLUM

TABLE I

Configuration	θ_a	θ_b	θ_n	Energy	Degeneracy
			Dense Phases (1×1)		
$\begin{array}{c} a \\ a\ a \end{array}$	1	0	0	$3\phi_{aa} + \mu_a$	1
$\begin{array}{c} b \\ b\ b \end{array}$	0	1	0	$3\phi_{bb} + \mu_b$	1
$\begin{array}{c} n \\ n\ n \end{array}$	0	0	1	0	1
			Diluted Phases $(\sqrt{3} \times \sqrt{3})$		
$\begin{array}{c} b \\ a\ a \end{array}$	$\frac{2}{3}$	$\frac{1}{3}$	0	$\phi_{ab} + (\frac{1}{2})\phi_{bb} + (\frac{3}{2})\phi_{aa}$ $+ (\frac{1}{3})\mu_b + (\frac{2}{3})\mu_a$	3
$\begin{array}{c} a \\ b\ b \end{array}$	$\frac{1}{3}$	$\frac{2}{3}$	0	$\phi_{ab} + (\frac{1}{2})\phi_{aa} + (\frac{3}{2})\phi_{bb}$ $+ (\frac{1}{3})\mu_a + (\frac{2}{3})\mu_b$	3
$\begin{array}{c} n \\ a\ a \end{array}$	$\frac{2}{3}$	0	$\frac{1}{3}$	$\phi_{aa} + (\frac{2}{3})\mu_a$	3
$\begin{array}{c} n \\ b\ b \end{array}$	0	$\frac{2}{3}$	$\frac{1}{3}$	$\phi_{bb} + (\frac{2}{3})\mu_b$	3
$\begin{array}{c} a \\ n\ n \end{array}$	$\frac{1}{3}$	0	$\frac{2}{3}$	$(\frac{1}{3})\mu_a$	3
$\begin{array}{c} b \\ n\ n \end{array}$	0	$\frac{1}{3}$	$\frac{2}{3}$	$(\frac{1}{3})\mu_b$	3
$\begin{array}{c} a \\ b\ n \end{array}$	$\frac{1}{3}$	$\frac{1}{3}$	$\frac{1}{3}$	$\phi_{ab} + (\frac{1}{3})\mu_a + (\frac{1}{3})\mu_b$	6

Then there are two possible situations: either the repulsive interactions predominate and

$$-\phi_{aa} > \phi_{ab} \quad \text{(case I)} \tag{139}$$

or the ab attractions predominate

$$-\phi_{aa} < \phi_{ab} \quad \text{(case II)}. \tag{140}$$

The phases in case I for a possible set of parameters is pictured in Fig. 7. Here we may have phases ranging from pure a for very large negative μ_a to

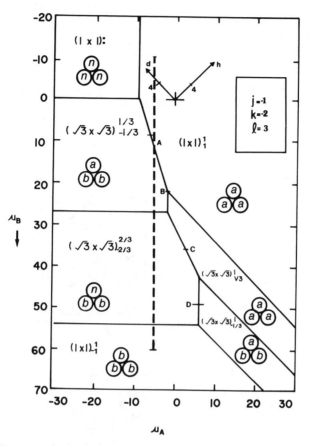

Figure 7. Phase diagram for the three-state adsorption model, case I (from Rikvold et al.[89]).

pure b for very large negative μ_b. In the example of case II the ab attraction predominates so as to render the dilute phases with $\theta_b = 1/3, 2/3$ thermodynamically unstable. This case is pictured in Fig. 8.

Complicated phase diagrams may exist in the neighborhood of the point of zero charge for systems in which there is strong chemisorption of some of the species. In the absence of specific adsorption one would expect that either μ_a or μ_b is large, but not both at the same time. In these cases we are in the asymptotic regime where the system can be treated by two-state models. This is done in the next section.

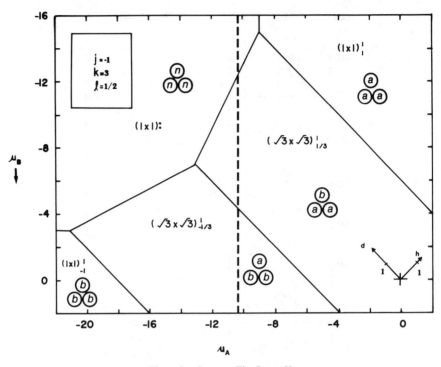

Figure 8.   Same as Fig. 7 case II.

C.   The Two-State Adsorption Model

We address now the case in which only one of the ionic species of the solution or the solvent is adsorbed. In this case the structure of the phase diagrams is simpler, which at the same time allows for a more general discussion, because we can include the cases in which the size of the adsorbate is variable.

The following discussion is taken from the work of Huckaby and Blum.[90] We consider a system in which we have a triangular lattice of sticky sites on a hard plane surface. The spacing between the sites of the lattice is d. This surface is in contact with a solution. Only two states of occupation are allowed: the sites are either occupied by an ion or by the solvent, or alternatively in the case of a pure fluid, by a fluid particle or none. The fluid particles have an exclusion diameter σ, which may or may not be associated with a hard core potential. Otherwise the interactions are arbitrary. We assume however that the pair correlations on the surface decay sufficiently fast that we need to take into account first neighbor interactions only. There are two possible situations. If the adsorbate diameter σ is smaller than the lattice spacing d, there are two possible phases, a dense, crystalline one and a dilute, disordered one. There is a first-order transition between them. If the adsorbed particles exclude all next-nearest neighbors,

$$d < \sigma < \sqrt{3}d, \tag{141}$$

and the problem is exactly analogous to the hard hexagon problem [93] of Baxter. In this case there is a second-order phase transition between an ordered $\sqrt{3} \times \sqrt{3}$ phase and a disordered one.

1. Single-Site Occupancy

Consider again Eq. (114) together with the superposition approximation Eq. (133):

$$Z/Z^0 = \sum_{\substack{\text{all sites} \\ \text{on the surface}}} \frac{\lambda^s \prod_{i=1}^{s} \rho^0(\mathbf{r}_i)}{s!} \exp\left[-\beta \sum_{nn} w_2^0(\mathbf{r}_i, \mathbf{r}_j)\right], \tag{142}$$

where the subscript nn means that the pairs $(\mathbf{r}_i, \mathbf{r}_j)$ in the sum inside the square brackets are nearest neighbors. As was done before (see Eq. 134), we can write Eq. (142) as a lattice gas partition function,[94]

$$Z/Z^0 = \sum_{\{t_i\}} \exp\left[-\beta w \sum_{nn} t_i t_j + \beta \mu \sum_i t_i\right] \qquad t_i = 0, 1, \tag{143}$$

where t_i is the occupation number of site i which can be either 0 for the site when it is occupied by the solvent (or empty) and 1 when it is occupied. Furthermore,

$$\beta w = -\ln g_2^0(d), \tag{144}$$

$$\beta \mu = \ln[\lambda \rho^0(0)]. \tag{145}$$

This partition function can be mapped onto an Ising model with spin variables $s_i = \pm 1$ by means of the transformation

$$s_i = 2t_i - 1 \qquad \text{or} \qquad t_i = (s_i + 1)/2 \tag{146}$$

In this case Eq. (143) becomes the partition function for the Ising model

$$Z/Z^0 = C \sum_{\{t_i\}} \exp\left[-\frac{\beta w}{4} \sum_{nn} (s_i s_j - 2s_i - 2s_j) + \frac{1}{2}\beta\mu \sum_i s_i\right] \qquad s_i = \pm 1. \tag{147}$$

where C is a constant that is irrelevant to our calculation and $w/4 = J$ is the Ising parameter, which is bigger than zero in the ferromagnetic case, and smaller than zero in the antiferromagnetic case. The variable that plays the role of the external magnetic field is h, given by

$$2h = \mu - wq/2, \tag{148}$$

where q is the coordination number of the lattice. In the case of the triangular lattice $q = 6$. The two-state ferromagnetic Ising model has a first-order phase transition when $h = 0$. From Eq. (148) this means

$$\mu = wq/2, \tag{149}$$

and using the definitions Eq. (144) and Eq. (145) we get the exact condition for phase transitions when $w > 0$

$$\mathcal{Z} = \lambda \rho^0(0) = [g_2^0(d)]^{-3}. \tag{150}$$

The preceding analysis can be illustrated by a fluid of hard spheres in contact with a sticky triangular lattice of spacings $d = \sigma$. In this case a good estimate of both the contact density and the pair distribution function are obtained from the Percus Yevick theory

$$\rho^0(0) = \rho \frac{(1 + 2\eta)}{(1 - \eta)^2}, \tag{151}$$

$$g_2^0(\sigma) = \frac{(1 + \eta/2)}{(1 - \eta)^2}, \tag{152}$$

$$\eta = (1/6)\pi \rho \sigma^3, \tag{153}$$

where η is the fraction of occupied volume. Replacing into Eq. (150) we get

$$\lambda = (1/6)\pi \sigma^3 \frac{(1 - \eta)^8}{\eta(1 + 2\eta)(1 + \eta/2)^3}. \tag{154}$$

This relation is a necessary but not sufficient condition for the occurrence of phase transitions.

A sufficient condition for the occurrence of a phase transition can be obtained from the classic work of Potts on the magnetization of the ferromagnetic Ising model on the triangular lattice.[95] The magnetization in the Ising model is the difference in the number of spins up ρ_u and spins down ρ_d,

$$I(x) = \rho_u - \rho_d \tag{155}$$

or using Eq. (146)

$$I(x) = 2\theta - 1, \tag{156}$$

where θ is the fraction of occupied sites. Following Potts we use the variable

$$x = e^{-\beta w/2} = 1/(g_2^0(d))^{1/2}. \tag{157}$$

The result for the spontaneous magnetization is

$$I(x) = \left[1 - \frac{16x^6}{(1 + 3x^2)(1 - x^2)^3}\right]^{1/8}. \tag{158}$$

Using Eqs. (156) and (157) yields the fraction of occupied sites as a function of the pair interaction function

$$\theta = \frac{1}{2} - \frac{1}{2}\left[1 - \frac{16g_2^0(d)}{(g_2^0(d) + 3)(g_2^0(d) - 1)^3}\right]^{1/8}. \tag{159}$$

The critical value of θ occurs when $\theta = 1/2$. Solving Eq. (159) for the contact pair correlation function yields

$$g_2^0(d) = 3. \tag{160}$$

From Eq. (150) we get the value for the critical sticky parameter λ

$$\lambda \rho^0(0) = 1/27. \tag{161}$$

In our example of hard core fluid in the Percus–Yevick approximation, we use these equations together with Eqs. (151) and (152) to get the critical value of the excluded volume fraction

$$\eta_{\text{crit}} = 0.3712 \tag{162}$$

and the sticky parameter

$$\lambda_{\text{crit}} = 0.01185\sigma^3. \tag{163}$$

This system undergoes a first-order phase transition. The isotherms for various values of the parameter λ are shown in Fig. 9. These exact results are

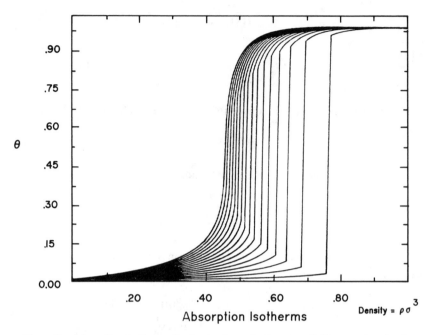

Figure 9. Adsorption isotherms for hard spheres in the mean field approximation for a triangular lattice of sticky sites.[84]

in qualitative agreement with the mean field theory of Badiali et al.,[84] where the first-order phase transition is also predicted. But the quantitative agreement is not good, which illustrates the pitfalls of mean field theory. In the computer simulations of Caillol et al.[96] the conditions for the occurrence of a first-order phase transition are not met, because the adsorption sites are of finite size, and for that reason the occupancy of a site may prevent nearest-neighbor occupation. In this case we expect a second-order phase transition to occur.

When there are longer-ranged interactions beyond the nearest-neighbor interactions, Dobrushin[97] has shown that the first-order phase transition still occurs. However, an exact relation such as Eq. (150) is not available.

As was mentioned already in the electrochemical case the contact pair correlation function of ions of equal sign is practically zero because of the Coulomb repulsion which prevents ions of equal sign to approach each other. However, condensed phases in the *ad* layers are observed in electrochemistry. In particular the underpotential deposition of some metals on electrodes occurs at certain very well-defined values of the potential bias.[98] For example, the deposition of Cu on the Au (Eq. 111) face forms two phases according to

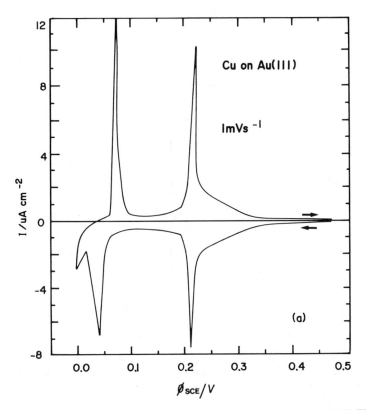

Figure 10. Cyclic potential curves for underpotential deposition of Cu on Au(111). The first peak (low-density phase), corresponds to a coverage $\theta = 0.6$. The second peak corresponds to a dense, commensurate *ad* layer.[99]

the deposition potential.[99] These phases have been observed ex situ[100] and in situ.[101] At a lower potential a dilute ordered $\sqrt{3} \times \sqrt{3}$ phase is formed. At a higher potential a dense commensurate phase is formed (see Fig. 10). It is clear from these considerations that in the dense *ad* layer case the ions must be discharged, because then they would form a metallic bond, which makes w positive and therefore ferromagnetic. This is supported by the features of the EXAFS spectra. In the high-density phase the near-edge structure corresponds to that of metallic copper, which has a characteristic double peak (Fig. 11). The dilute $\sqrt{3} \times \sqrt{3}$ phase has the white line characteristic of the charged ions. We may assume then that in this case the Cu retains part of its charge, so that the interactions are in this case repulsive, which corresponds to the antiferromagnetic case.

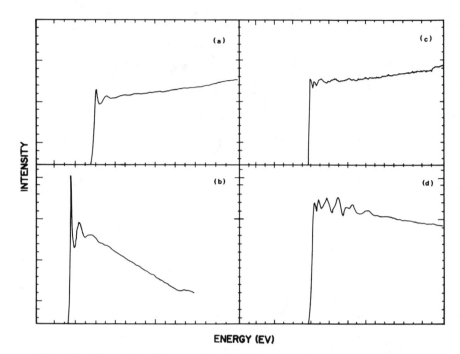

INTENSITY

ENERGY (EV)

Figure 11. Comparison of the near-edge adsorption peaks for the underpotential deposited Cu on Au (Eq. 111). (*a*) Spectra of the low-density phase. The near-edge structure has a single high peak, similar to that of $CuSO_4$, shown in (*b*). The high-density phase spectrum (*c*) has a double peak, similar to that of Cu foil (*d*).

2. Multiple-Site Occupancy: The Hard Hexagon Case

If the occupation of one site in the triangular lattice also excludes the nearest neighbors, the problem is equivalent to the hard hexagon problem of Baxter.[87] This problem can be solved when the interactions between the hard hexagons are neglected.

The thermodynamics of the hard hexagon model was recently worked out by Joyce.[102] In terms of the lattice fugacity

$$\mathscr{L} = \lambda \rho^0(0). \tag{130}$$

The system undergoes a second-order phase transition between an ordered solid-like phase and a disordered one. The transition occurs when

$$\mathscr{L}_{\text{crit}} = (11 + 5\sqrt{5})/2 = 11.09. \tag{164}$$

The fraction of occupied sites is

$$\theta_{\text{crit}} = (5 - \sqrt{5})/10 = 0.2764. \tag{165}$$

We remark that in the limit of highest possible density the occupied sites fraction is $\theta = 1/3$.

For the low-density phase the equation of state is

$$\mathcal{L} = \frac{1}{4\theta^6(1-\theta)} [Q_0^2 Q_1^{1/2} + Q_2 - Q_0(2Q_3 + 2Q_2 Q_1^{1/2})^{1/2}], \tag{166}$$

$$Q_0 = 1 - 5\theta + 5\theta^2,$$

$$Q_1 = Q_0(1 - \theta + \theta^2),$$

$$Q_2 = (1 - 2\theta)(1 - 11\theta + 44\theta^2 - 77\theta^3 + 66\theta^4 - 33\theta^5 + 11\theta^6),$$

$$Q_3 = 1 - 16\theta + 106\theta^2 - 378\theta^3 + 803\theta^4 - 1080\theta^5 + 962\theta^6$$
$$- 576\theta^7 + 219\theta^8 - 50\theta^9 + 10\theta^{10},$$

where for the ordered phase

$$\mathcal{L} = \frac{2(2-3\theta)(1-\theta)^3}{(-1 + 12\theta - 45\theta^2 + 66\theta^3 + 33\theta^4 - ((1 - 5\theta + 5\theta^2)^3(1 - 9\theta + 9\theta^2))^{1/2})}. \tag{167}$$

Consider the case of an electrolyte in the neighborhood of a charged electrode. The simplest model is the restricted primitive model of a continuum dielectric and a smooth imageless surface. The contact probability is computed from the Gouy–Chapman theory discussed in Section I:

$$\rho_i(0)/\rho_i = \exp[-z_i\psi_i(0)], \tag{168}$$

where we are using Eq. (7) together with the definition

$$\psi_i(0) = \beta e\phi_i(0), \tag{169}$$

we get

$$\mathcal{L} = \lambda\rho_i \exp[-z_i\psi_i(0)], \tag{170}$$

where $\psi_i(0)$ is the reduced potential at the inner Helmholtz plane. Equating Eq. (170) to either Eq. (166) or Eq. (167) produces a suggestive form of an adsorption isotherm that is shown in Fig. 12.

From Eqs. (164) and (170) we deduce that in reduced units there is a critical

218   L. BLUM

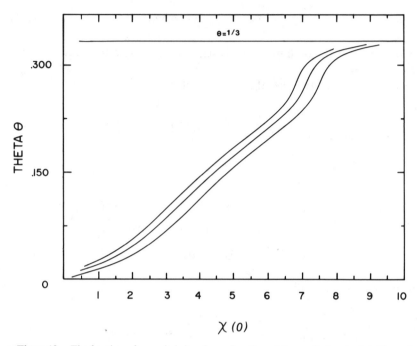

Figure 12. The fraction of occupied sites θ as a function of the applied potential. The units are adimensional. There a second-order phase transition at $\theta = 0.2764$.

potential at which a phase transition occurs between an ordered and a disordered phase,

$$z_i \psi_{crit}(0) = \ln[\lambda \rho_i] - 2.406. \qquad (171)$$

At this point a second-order phase transition occurs in the surface, which is also seen in the adsorption isotherm.

Acknowledgments

The author is indebted to Dr. D. A. Huckaby and Dr. J. G. Gordon for useful discussions. This work was supported by the Office of Naval Research.

References

1. A. T. Hubbard, *Crit. Rev. Anal. Chem.* **3**, 201 (1973). E. J. Yaeger, *Electroanal. Chem.* **150**, 181, 535 (1983). D. M. Kolb and W. N. Hansen, *Surface Sci.* **101**, 109 (1979). D. M. Kolb, *Ber. Bunsenges. Phys. Chem.* **92**, 1175 (1988). K. Yamamoto, D. M. Kolb, R. Koetz, and J. Lehmpful, *J. Electroanal. Chem.* **96**, 233 (1979) and references cited therein. G. N. Salaita, F. Lu, L. Laguren-Davidson, and A. T. Hubbard, *J. Electroanal. Chem.* **229** (1987). D. Aberdam, R. Durand, R. Faure, and F. El-Omar, *Surface Sci.* **171** 303, (1986).

2. For a recent book see B. K. Teo, *EXAFS: Basic Principles and Data Analysis*, Springer-Verlag, Berlin, 1986.

3. L. Blum, H. D. Abruna, J. H. White, M. J. Albarelli, J. G. Gordon, G. L. Borges, M. G. Samant, and O. R. Melroy, *J. Chem. Phys.* 85, 6732 (1986). M. G. Samant, G. L. Borges, J. G. Gordon, L. Blum, and O. R. Melroy, *J. Am. Chem. Soc.* 105, 5970 (1987). H. D. Abruna, J. H. White, M. J. Albarelli, G. M. Bommarito, M. J. Bedzyk, and M. McMillan, *J. Chem. Phys.* 92, 7045 (1988).

4. W. C. Marra, P. Eisenberger, and A. Y. Choh, *J. Appl. Phys.* 50, 6972 (1979). I. K. Robinson, *Phys. Revs.* B33, 3830 (1986).

5. M. G. Samant, M. F. Toney, G. L. Borges, L. Blum, and O. R. Melroy, *J. Phys. Chem.* 92, 220 (1988). M. G. Samant, M. F. Toney, G. L. Borges, L. Blum, and O. R. Melroy, *Surf. Sci.* 193, L29 (1988). M. Fleischmann, P. J. Hendra, and J. Robinson, *Nature*, 288, 152 (1980). M. Fleischmann and B. W. Mao, *J. Electroanal. Chem.* 229, 125 (1987); 247, 297, 311 (1988).

6. B. Batterman, *Phys. Revs.* 133, A759 (1964). J. A. Golovchenko, B. Batterman, and W. L. Brown, *Phys. Revs.* B10, 4239 (1974).

7. G. Materlik, J. Zegerhagen, and W. Uelhoff, *Phys. Revs.* B32, 5502 (1985). G. Materlik, M. Schmah, J. Zegerhagen, and W. Uelhoff, *Ber. Bunsenges. Phys. Chem.* 91, 292 (1987). M. Bedzyk, D. Bilderback, J. H. White, H. D. Abruna, and G. M. Bommarito, *J. Phys. Chem.* 90, 4926 (1986). M. Bedzyk, D. Bilderback, G. M. Bommarito, M. Caffrey, and J. S. Schildkraut, *Science* 241, 1788 (1988). B. M. Ocko, *Nato Conference Proceedings*, B. Guiterrez and C. Melendres (Eds.), 1988.

8. R. J. Gale, *Spectroelectrochemistry, Theory and Practice*, Plenum, New York, 1988.

9. M. Fleischmann, J. Oliver, and J. Robinson, *Electrochimica Acta* 31, 899 (1986). R. K. Chang and T. E. Furtak, *Surface Enhanced Raman Scattering*, Plenum, New York, 1985.

10. A. Bewick, and S. Pons, in *Advances in Infrared and Raman Spectroscopy*, Vol. 12, R. J. H. Clark and R. E. Hester (Eds.), Wiley, New York, 1985.

11. G. L. Richmond, J. M. Robinson, and V. L. Shannon, *Prog. Suf. Sci.* 28, 1 (1988).

12. O. Lev, F. R. Fan, and A. J. Bard, *J. Electrochem. Soc.* 135, 783 (1988). R. Sonnenfeld, and B. C. Schardt, *Appl. Phys. Lett.* 49, 1172 (1986). K. Itaya and E. Tomita, *Surf. Sci.* 201, L507 (1988). P. Lustenberger, H. Rohrer, R. Christoph, and J. Siegenthaler, *J. Electroanal Chem.* 243, 225 (1988). J. Wiechers, T. Twomey, D. M. Kolb, and R. J. Behm, *J. Electroanal Chem.* 248, 451 (1988). M. H. S. Hottenhuis, M. A. H. Mickers, J. W. Gerntsen, and J. P. van der Eerden, *Surf. Sci.* 206, 259 (1988).

13. D. M. Kolb, *Ber. Bunsenges., Phys. Chem.* 92, 1175 (1988).

14. O. R. Melroy, K. Kanazawa, J. G. Gordon, and D. Buttry, *Langmuir*, 2, 697 (1986). R. Schumacher, G. L. Borges, and K. Kanazawa, *Surf. Sci.* 163, L621 (1985). R. Schumacher and W. Stoeckel, *Ber. Bunsenges. Phys. Chem.* 93, 600 (1989).

15. J. W. Schultze and K. J. Vetter, *Ber. Bunsenges. Phys. Chem.* 76, 920 (1972).

16. J. N. Israelachvili, *Chem. Scr.* 25, 7 (1985). R. Horn and J. N. Israelachvili, *J. Chem. Phys.* 75, 1400 (1981). H. K. Christenson and R. Horn, *Chem. Scr.* 25, 37 (1985). J. N. Israelachvili and P. M. McGuiggan, *Science* 241, 795 (1988).

17. R. Kotz, D. M. Kolb, and J. K. Sass, *Surface Sci.* 69, 359 (1977). D. M. Kolb, W. Boeck, and S. H. Liu, *Phys. Revs. Lett.* 47, 1921 (1981).

18. V. Russier, *Surface Sci.* 214, 304 (1989) and references contained therein.

19. D. M. Kolb, W. Boeck, K. M. Ho, and S. H. Liu, *Phys. Revs. Lett.* 47, 1921 (1981). W. Boeck and D. M. Kolb, *Surface Sci.* 118, 613 (1982). J. Schneider, C. Franke, and D. M. Kolb, *Surface Sci.* 198, 277 (1988).

20. S. Levine and R. W. Fawcett, *J. Electroanal. Chem.* 99, 265 (1979). R. Parsons and F. R. G. Zobel, *J. Electroanal. Chem.* 9, 333 (1965). A. Hamelin, in *Modern Aspects of Electrochemistry*, Vol. 16, B. E. Conway, R. E. White, and J. O'M. Bockris (Eds.), Plenum, New York, 1985.

21. A. Hamelin, L. Stoicoviciu, L. Doubova, and S. Trasatti, *Surface Sci.* **201**, L498 (1988).
22. B. Jancovici, *Phys. Revs. Lett.* **46**, 386 (1981). A. Alastuey and J. L. Lebowitz, J. Phys. (France), **45**, 1859, (1984). M. L. Rosinberg and L. Blum, *J. Chem. Phys.*, **81**, 3700 (1984). M. L. Rosinberg, J. L. Lebowitz, and L. Blum, *J. Stat. Phys.*, **44**, 153 (1986). F. Cornu and B. Jancovici, *J. Chem. Phys.*, **90**, 2444 (1989).
23. F. Vericat and L. Blum, *J. Chem. Phys.* **82**, 1492 (1985). F. Vericat and L. Blum, *J. Stat. Phys.* (submitted).
24. C. Y. Lee, J. A. McCammon, and P. J. Rossky, *J. Chem. Phys.* **80**, 4448 (1984). J. P. Valleau and A. A. Gardner, *J. Chem. Phys.* **86**, 4162, 4171 (1987). E. Spohr and K. Heinziger, *Chem. Phys. Lett.* **123**, 218 (1986). E. Spohr, *J. Phys. Chem.* **93**, 6171 (1989). N. G. Parsonage and D. Nicholson, *J. Chem. Soc. Faraday Trans.* **82**, 1521 (1986); **83**, 663, (1987).
25. S. L. Carnie, D. Y. C. Chan, and G. R. Walker, *Mol. Phys.* **43**, 1115 (1981).
26. D. Bratko, L. Blum, and A. Luzar, *J. Chem. Phys.* **83**, 6367 (1985). P. T. Cummings and L. Blum, *J. Chem. Phys.* **84**, 1833 (1986). L. Blum, P. T. Cummings, and D. Bratko, *J. Chem. Phys.*, in press.
27. S. Levine and G. M. Bell, *Discuss. Faraday Soc.* **42**, 69 (1966). S. Levine and C. W. Outhwaite, *J. Chem. Soc. Faraday II* **74**, 1670 (1978). C. W. Outhwaite, *J. Chem. Soc. Faraday II* **74**, 1214 (1978). L. B. Bhuiyan, C. W. Outhwaite, and S. Levine, *Mol. Phys.* **42**, 1271 (1981). C. W. Outhwaite, L. B. Bhuiyan, and S. Levine, *Chem. Phys. Lett.* **78**, 413 (1981). C. W. Outhwaite and L. B. Bhuiyan, *J. Chem. Soc. Faraday II* **79**, 707 (1983). C. W. Outhwaite and L. B. Bhuiyan *J. Chem. Phys.* **85**, 4206 (1986).
28. R. Evans and T. Sluckin, *Mol. Phys.* **40**, 413 (1980).
29. S. L. Carnie and G. M. Torrie, *Adv. Chem. Phys.* **56**, 141 (1984).
30. D. Henderson, F. F. Abraham, and J. A. Barker, *Mol. Phys.* **31**, 1291 (1976).
31. M. Born and H. S. Green, *Proc. Roy. Soc. London,* **1988**, 10 (1946) J. Yvon, *La Theorie Statistique des Fluides*, Paris (1935). Dunod
32. R. Lovett, C. Y. Mou, and F. P. Buff, *J. Chem. Phys.* **65**, 2377 (1976). M. S. Wertheim, *J. Chem. Phys.* **65**, 2377 (1976).
33. J. G. Kirkwood and E. Monroe, *J. Chem. Phys.* **9**, 514 (1941).
34. G. M. Torrie and J. P. Valleau, *Chem. Phys. Lett.* **65**, 343 (1979); *J. Chem. Phys.* **73**, 5807 (1980). G. M. Torrie, J. P. Valleau, and G. N. Patey, *J. Chem. Phys.* **73**, 5807 (1982). G. M. Torrie and J. P. Valleau, *J. Chem. Phys.* **76**, 4615 (1982). G. M. Torrie and J. P. Valleau, *J. Phys. Chem.* **86**, 3251 (1982). J. P. Valleau and G. M. Torrie, *J. Chem. Phys.* **81**, 6291 (1984). G. M. Torrie, J. P. Valleau, and C. W. Outhwaite, *J. Chem. Phys.* **81**, 6296 (1984).
35. S. W. de Leeuw, J. W. Perram, and E. R. Smith, *Proc. Roy. Soc.* **A373**, 27, 57 (1980); **A388**, 177 (1983); *Ann. Revs. Phys. Chem.*, **37**, 245 (1986).
36. G. Gouy, *J. Phys.* **9**, 457 (1910). D. L. Chapman, *Phil. Mag.* **25**, 475 (1913).
37. O. Stern, *Z. Elektrochem.* **30**, 508 (1924).
38. J. P. Valleau and G. M. Torrie, *J. Chem. Phys.* **76**, 4623 (1982).
39. L. B. Bhuiyan, L. Blum, and D. Henderson, *J. Chem. Phys.* **78**, 1902 (1983).
40. See, for example, I. S. Gradshtein and I. M. Rhyzik, *Tables of Integrals, Series and Products,* Academic Press, New York, 1965. D. Grahame, *J. Chem. Phys.* **21**, 1054 (1953).
41. J. P. Valleau and G. M. Torrie, *J. Chem. Phys.* **76**, 4623 (1982). L. B. Bhuiyan, L. Blum, and D. J. Henderson, *J. Chem. Phys.* **78**, 442 (1983).
42. L. Blum, J. Hernando, and J. L. Lebowitz, *J. Phys. Chem.* **87**, 2825 (1983).
43. J. L. Lebowitz and J. K. Percus, *J. Math. Phys.* **4**, 116, 248 (1963).
44. F. P. Buff and F. H. Stillinger, *J. Chem. Phys.* **25**, 312 (1956). F. H. Stillinger and F. P. Buff, *J. Chem. Phys.* **37**, 1 (1962).
45. G. Stell, *Physica,* **29**, 517 (1963).
46. J. P. Hansen and I. R. McDonald, *Theory of Simple Liquids,* Academic Press, New York, 1976.

47. D. Henderson, F. F. Abraham, and J. A. Barker, *Mol. Phys.* **31**, 1291 (1976). J. W. Perram and E. R. Smith, *Chem. Phys. Lett.* **39**, 328 (1976).
48. L. Blum and G. Stell, *J. Stat. Phys.* **15**, 439 (1976). D. E. Sullivan and G. Stell, *J. Chem. Phys.* **67**, 2567 (1977).
49. D. Henderson and L. Blum, *J. Chem. Phys.* **69**, 5441 (1978).
50. S. L. Carnie and D. Y. C. Chan, *Chem. Phys. Lett.* **77**, 437 (1981).
51. L. Blum, J. L. Lebowitz, and D. Henderson, *J. Chem. Phys.* **72**, 4249 (1980).
52. D. Henderson, L. Blum, and W. R. Smith, *Chem. Phys. Lett.* **63**, 381 (1979).
53. L. Blum and D. Henderson, *J. Electroanal. Chem.* **111**, 217 (1980). S. L. Carnie, D. Y. C. Chan, D. J. Mitchell, and B. Ninham, *J. Chem. Phys.* **74**, 1472 (1981). M. Lozada-Cassou, R. Saavedra, and D. Henderson, *J. Chem. Phys.* **77**, 5150 (1982). M. Lozada-Cassou and D. Henderson, *J. Phys. Chem.* **87**, 2821 (1982). J. Barojas, M. Lozada-Cassou, and D. Henderson, *J. Phys. Chem.* **87**, 4547 (1982). S. L. Carnie, *Mol. Phys.* **54**, 509 (1985).
54. P. Ballone, G. Pastore, and M. P. Tosi, *J. Chem. Phys.* **85**, 2943 (1986).
55. Y. Rosenfeld and L. Blum, *J. Chem. Phys.* **85**, 2197 (1986).
56. P. Nielaba and F. Forstmann, *Chem. Phys. Lett.* **117**, 46 (1985).
57. M. S. Wertheim, *J. Chem. Phys.* **65**, 2377 (1976).
58. T. L. Croxton and D. A. McQuarrie, *Mol. Phys.* **42**, 141 (1981).
59. C. Caccamo, G. Pizzimenti, and L. Blum, *J. Chem. Phys.* **84** 3327 (1986). E. Bruno, C. Caccamo, and G. Pizzimenti, *J. Chem. Phys.* **86**, 5101 (1987).
60. S. L. Carnie and D. Y. C. Chan, *Mol. Phys.* **51**, 1047 (1984).
61. M. Plischke and D. Henderson, *J. Phys. Chem.* **88**, 2712 (1988).
62. D. Henderson and M. Plischke, *Mol. Phys.* **62**, 801 (1987).
63. R. M. Nieminen and N. W. Ashcroft, *Phys. Revs.* **A24**, 560 (1981). S. M. Foiles and N. W. Ashcroft, *Phys. Revs.* **B25**, 1366 (1982).
64. R. Lovett, C. Y. Mou, and F. P. Buff, *J. Chem. Phys.* **65**, 2377 (1976).
65. A. R. Altenberger, *J. Chem. Phys.* **76**, 1473 (1982).
66. M. Plischke and D. Henderson, *J. Chem. Phys.* **90**, 5738 (1989). D. Henderson and M. Plischke, *J. Phys. Chem.* **92**, 7177 (1988).
67. P. J. Colmenares and W. Olivares, *J. Chem. Phys.* **90**, 1977 (1986). P. J. Colmenares and W. Olivares, *J. Chem. Phys.* **88**, 3221 (1988).
68. R. Kjellander and S. Marcelja, *Chem. Phys. Lett.* **112**, 49 (1984); *J. Chem. Phys.* **82**, 2122 (1985); *Chem. Phys. Lett.* **127**, 402 (1986).
69. D. J. Henderson, L. Blum, and J. L. Lebowitz, *J. Electroanal. Chem.* **102**, 315 (1979).
70. L. Blum and D. J. Henderson, *J. Chem. Phys.* **74**, 1902 (1981).
71. L. Blum, M. L. Rosinberg, and J. P. Badiali, *J. Chem. Phys.* **90**, 1285 (1989).
72. F. Cornu, *J. Stat. Phys.* **54**, 681 (1989).
73. L. Blum, Ch. Gruber, J. L. Lebowitz, and Ph. A. Martin, *Phys. Revs. Lett.* **48**, 1769 (1982). Ch. Gruber, J. L. Lebowitz, and Ph. A. Martin, *J. Chem. Phys.* **75**, 944 (1981).
74. B. Jancovici, *J. Stat. Phys.* **28**, 43 (1982).
75. A. Alastuey and Ph. A. Martin, *Europhys. Lett.* **6**, 385 (1988).
76. F. H. Stillinger and R. Lovett, *J. Chem. Phys.* **48**, 3858, **49**, 1991 (1968).
77. C. W. Outhwaite, *Chem. Phys. Lett.* **24**, 73 (1974).
78. S. L. Carnie and D. Y. C. Chan, *Chem. Phys. Lett.* **77**, 437 (1981).
79. Ph. A. Martin, *Revs. Mod. Phys.* **60**, 1075 (1988).
80. L. Blum, D. Henderson, J. L. Lebowitz, Ch. Gruber, and Ph. A. Martin, *J. Chem. Phys.* **75**, 5974 (1981).
81. D. G. Triezenberg and R. Zwanzig, *Phys. Revs. Lett.* **28**, 1183 (1972).
82. J. R. Henderson and F. van Swol, *Mol. Phys.* **51**, 991 (1984).
83. M. L. Rosinberg, J. L. Lebowitz, and L. Blum, *J. Stat. Phys.* **44**, 153 (1986).
84. J. P. Badiali, L. Blum, and M. L. Rosinberg, *Chem. Phys. Lett.* **129**, 149 (1986).

85. L. Boltzmann, *Vorlesungen ueber Gastheorie*, Leipzig, 1912.

86. I. Langmuir, *J. Am. Chem. Soc.* **39**, 1848 (1917).

87. R. J. Baxter, *J. Chem. Phys.* **49**, 2770 (1968). J. W. Perram and E. R. Smith, *Proc. Roy. Soc.* **A353**, 193 (1977).

88. R. J. Baxter, *Exactly Solved Problems in Statistical Mechanics*, Academic Press, New York, 1982.

89. P. A. Rikvold, J. B. Collins, G. D. Hansen, and J. D. Gunton, *Surf. Sci.* **203**, 501 (1988).

90. D. A. Huckaby and L. Blum, *J. Chem. Phys.*, **92**, 2646 (1990).

91. J. B. Collins, P. A. Rikvold, and E. T. Gawlinski, *Phys. Revs.* **B38**, 6741 (1988).

92. Y. Saito, *J. Chem. Phys.* **74**, 713 (1981).

93. R. J. Baxter, *J. Stat. Phys.* **26**, 427 (1981).

94. T. D. Lee and C. N. Yang, *Phys. Revs.* **87**, 410 (1950).

95. R. B. Potts, *Phys. Revs.* **88**, 352 (1952).

96. J. M. Caillol, D. Levesque, and J. J. Weis, *J. Chem. Phys.* **87**, 6150 (1987).

97. R. L. Dobrushin, *Funct. Anal. Appl.* **2**, 302 (1952).

98. L. Blum, H. D. Abruna, J. H. White, M. J. Albarelli, J. G. Gordon, G. L. Borges, M. G. Samant, O. R. Melroy, *J. Chem. Phys.* **85**, 6732 (1986).

99. D. M. Kolb, K. Al Jaaf-Golze, and M. S. Zei, *DECHEMA Monographien*, **12**, 53, Verlag Chemie Weinheim (1986).

100. M. Zei, G. Qiao, G. Lehmpful, and D. M. Kolb, *Ber. Bunsenges. Phys. Chem.* **91**, 349 (1987).

101. O. R. Melroy, M. G. Samant, G. L. Borges, J. G. Gordon, L. Blum, J. H. White, M. J. Albarelli, M. McMillan, H. D. Abruna, *Langmuir*, **4**, 728 (1988).

102. G. S. Joyce, *J. Phys. A: Math. Gen.* **21**, L983 (1988).

DYNAMICS OF DENSE POLYMER SYSTEMS: COMPUTER SIMULATIONS AND ANALYTIC THEORIES

JEFFREY SKOLNICK

Molecular Biology Department, Research Institute of Scripps Clinic, La Jolla, California

and

ANDRZEJ KOLINSKI

Department of Chemistry, University of Warsaw, Warsaw, Poland

CONTENTS

I. INTRODUCTION

The elucidation of the mechanism by which an individual polymer chain moves in a concentrated solution or in a polymer melt (a viscoelastic liquid in which all the constituent molecules are polymers) has been among the central problems in polymer physics for over 40 years.[1,2] In addition to the complications arising from the effects of internal excluded volume on chain dynamics, there is the additional complexity in a melt due to the noncrossing constraint between the individual polymer chains. The problem of polymer melt dynamics is of interest not only in its own right as an example of an extremely complicated many-body problem, but also because its solution would have practical applications to the areas of polymer flow rheology, polymer adhesion, and polymer failure.[1] Thus considerable effort has been expended over the years to develop an effective single-particle picture capable of describing the dynamics of the chains in the melt. In this chapter we shall review a number of theoretical approaches to the solution of this problem.[3-22] Because of its inherent complexity, computer simulation techniques[20-24] and analytic theory[3-22] have been applied to explore the dynamics of entangled polymer systems.

A. Experimental Phenemonology

Before presenting an overview of the various theories that have been applied to treat melt dynamics, it is appropriate to summarize the salient experimental phemonenology that any successful theory must ultimately rationalize and encompass. Among the important characteristics of any given polymer property is the dependence of the property on molecular weight M (equivalently the degree of polymerization, n). In many cases, the absolute magnitude of a given property has proven extremely difficult to calculate, and the elucidation of the scaling of this property with M has been a major focus of various theories.

For a polymer melt composed of linear chains, the center of mass diffusion coefficient scales with n as[1,25-32]

$$D \sim n^{-1} \text{ when } n < n_{c'}, \tag{1a}$$

and with a further increase in n,

$$D \sim n^{-2} \text{ when } n > n_{c'}, \tag{1b}$$

where $n_{c'}$ is a crossover value of the degree of polymerization. However, more recent experiments have called into question whether the scaling embodied in Eq. (1b) is the asymptotic behavior.[33] Substantially stronger scaling with molecular weight has been reported[33], and in the case of concentrated polymer solutions Phillies[34] has found that a stretched exponential of the form

$$D = D_0 \exp(-\alpha c^{\nu}) \tag{1c}$$

fits experimental data extremely well. Here D_0 is the diffusion constant of the polymer at infinite dilution which scales as M^{-b} with b in the range 0.5–0.55, $\alpha \sim M$, and ν is molecular weight dependent crossing over from $\nu = 1.0$ to $\nu = 1/2$ as M increases.[11,12] Thus the question of the molecular weight dependence of D is not resolved.

Another transport property that has been extensively studied is the zero frequency shear viscosity η, which depends on n as follows:[3,26]

$$\eta \sim n^1 \text{ when } n < n_c \tag{2a}$$

and

$$\eta \sim n^{3.4} \text{ when } n > n_c. \tag{2b}$$

Observe that the value of the crossover degree of polymerization for viscosity and diffusion are unequal. Typical experimental data give n_c' of about $5n_c$.[35] Why the crossover values are different for the viscosity and the self-diffusion coefficient is not understood.

For a number of theoretical reasons[1,35] it has been questioned whether Eq. (2b) is the limiting power law behavior of η. A range of exponents of η has been reported, but the values tend to lie in the range 3.3–3.7, with 3.4 being the most prevalent.[3,35] A major conceptual difficulty with the 3.4 power law dependence arises from the following: In a polymer melt, the static excluded volume effect giving rise to chain expansion is screened and the mean square radius of gyration $\langle S^2 \rangle$ scales as $n^{1.0}$.[1,36] As pointed out by Colby et al.[35] the time required for molecules to diffuse a distance on the order of the radius of gyration is

$$t_d = \langle S^2 \rangle / D. \tag{3a}$$

Thus $t_d \sim n^{3.0}$ if Eq. (1b) holds.

On the other hand, the time scale for stress relaxation,

$$\tau_0 = J_e^0 \eta \tag{3b}$$

where J_e^0 is the recoverable shear compliance (a molecular weight independent quantity in the limit of large $M^{1.26}$). Hence, taking Eqs. (2b) and (3b) together implies that $\tau_0 \sim n^{3.4}$. In other words, if $\tau_0 \sim n^{3.4}$, then chains will move many radii of gyration before they undergo orientational relaxation. This apparently "nonphysical" result, when juxtaposed with the predictions of the reptation model of polymer melt dynamics, has led to the belief that in the asymptotic limit η should scale as $n^{3.0}$. Thus far, only one experiment by Colby et al.[35] on polybutadiene claims to find this n^3 behavior. However, these series of measurements, while clearly an experimental tour de force, have not been universally accepted as providing convincing evidence for the $n^{3.0}$ power law behavior of n.

Whatever the final resolution of this controversy, it is clear that as the length of the chains in a polymer melt increases, the behavior of the system changes drastically. The problem remains to identify the cause of this change in behavior; a reasonable qualitative explanation is that as the size of the chains increase, chain entanglements of some kind become important.

Thus far we have discussed the behavior of a melt of linear chains, which are by far the most extensively studied. However, rings are of intense interest because of the reptation model of polymer melt dynamics which asserts that for distances on the order of $\langle S^2 \rangle^{1/2}$, the other chains act to confine the chain of interest to a tube.[4-8,37] Thus, the dominant motion of the chain of interest involves the slithering down the tube, defined at zero time. In the reptation model, the motion of the ends is extremely important. Rings being entirely devoid of chain ends should, therefore, move substantially slower.[38] Experimentally, this doesn't appear to be the case. Melts of polymer rings have a lower viscosity and appear to be less entangled than the corresponding linear chains.[39,40] Roovers[40] has reported a scaling of η with n that is consistent with Eq. (2b); however, the absolute magnitude of η is less. Admittedly, due to synthetic difficulties, the range of molecular weights measured is substantially smaller than for linear chain melts; nevertheless, the rings that have been studied do crossover into what is considered the entangled regime; i.e., the regime where the polymeric nature of the medium is influencing the behavior of the chains.

B. Entanglements: An Overview

What is immediately apparent from the overview given above of the experimental phemonenology is that as the length of the chains comprising the melt increases, the chains behave differently; the origin of these effects has been ascribed to entanglements of some sort between the various chains.

What is the nature of these entanglements? One picture that comes to mind is that the chains form knots. Thus, one chain is trapped by a knot formed by another chain until one of the two ends comes through, and therefore, disengages the entanglement.[6,18] If this is true, the question arises as to how mobile the entanglements are. One might envision that these slip knots are basically immobile, for times, on the order of the τ_0, and if so the knots could therefore form the tube conjectured by reptation theory. Another possibility is that, in fact, the knots are not rigidly held in space but are quite mobile.[18] Thus the chains are not confined to a tube. Rather, the internal configurational relaxation can occur relatively rapidly, but the disengagement of the chain from the knots is the rate-determining step.

There is an alternative view of entanglements.[9,14,16] This viewpoint, while conceding that the disengagement of knots will contribute to the relaxation, holds that this isn't necessarily the dominant process. Rather, a chain need not be confined by knots to be dynamically entangled—looping of one chain around the other can work just as well. That this effect can be important is motivated by the analogy with tangled fishing line. Topologically, in tangled fishing line, there are no knots whatsoever, yet one loop gets trapped by another loop until there is a tangled mess. Thus entanglements are viewed to be intrinsically dynamic in nature, with the surrounding chain environment at intermediate times (corresponding to distances on the order of $\langle S^2 \rangle^{1/2}$) being viewed as quite fluid, much like in a small molecule liquid.

Unfortunately, experiment does not get at the exact nature of the entanglements. Based on estimates of the crossover behavior of the viscosity, whatever they are, entanglements are rare, occurring on the order of a hundred monomer units or so.[3] Computer simulations are an especially powerful technique for elucidating, at least in a qualitative sense, the microscopic mechanism of various physical processes;[20] the nature of the motion in a polymer melt in not an exception. However, the simulation of polymer melt dynamics is not a trivial matter. One must simulate a sufficiently large number of long chains for long times in order that entangled behavior be exhibited.[23] As discussed in further detail below, this is a nontrivial task. Here we point out that the simulations scale at least as the fourth power of n.

C. Rouse Model

One of the most surprising consequences of the experimental observations embodied in Eqs. (1) and (2) is that low molecular weight melts behave like Rouse chains, which is the simplest model for the dynamics of a polymer chain at infinite dilution.[41] In this model hydrodynamic interactions (the perturbation of the solvent flow surrounding one monomer due to the presence of the other monomers) are ignored, as are excluded volume effects.[41,42]

In what follows, the behavior of a Rouse chain will be required; we summarize here the salient features of the model. Qualitatively, the motion of a chain composed of identical monomers is isotropic at all times. For a chain composed of n monomers, the self-diffusion coefficient constant D_{Rouse} scales like n^{-1}. Furthermore, the mean-square displacement of the center of mass, $g_{\text{cm}}(t)$, is related to the self-diffusion coefficient, at all times t, by

$$g_{\text{cm}}(t) = 6Dt. \tag{4}$$

In the limit of very long chain lengths, the average mean square displacement of a bead exhibits the following behavior as a function of time:

$$g(t) \sim t^{1/2} \qquad t < \tau_{\text{Rouse}}, \tag{5a}$$

where τ_{Rouse} is the longest internal relaxation time and describes the decay of the end-to-end vector. τ_{Rouse} scales like $n^{2.0}$ if excluded volume effects are neglected. For distances such that $g(t)$ is appreciably greater than $2\langle S^2 \rangle$,

$$g(t) = 2\langle S^2 \rangle + 6Dt. \tag{5b}$$

Finally, the zero frequency shear viscosity scales like τ_{Rouse}/n, and thus the scaling behavior of Eq. (2a) is recovered.

D. Reptation Model

While the Rouse model has proven to be remarkably successful at characterizing the low molecular weight behavior of polymer melts, it cannot account for the enhanced dependence on molecular weight of D and n as the length of the chains increase.[25-34] The first class of models that proved capable of almost reproducing the molecular weight dependence of *both* the diffusion constant and the zero frequency shear viscosity is the reptation model.[4-8] It was originally proposed by de Gennes[4] to describe dynamics of a polymer chain moving in a gel, and later was fully developed by Doi and Edwards to apply to a polymer melt.[5-8] Since this is a widely accepted model against which all alternative theories have had to compete, it is appropriate here to summarize its features. We focus here on its application to linear chains. In the original reptation model,[1,4-8,37] one imagines that the surrounding chains produce entanglements that remain *static* for times on the order of the longest internal relaxation time. The beauty of this approach is that an extremely complicated many-body picture is reduced to a very simple single particle picture, namely, the dynamics of an isolated chain confined to a tortuous (Gaussian) tube. Lateral motions of the chain are always extremely limited, and the only way the chain can move distances on the order of the radius of gyration is by slithering out the ends of the tube, in a snake-like motion. The assertion that the dominant long wavelength motion is longitudinal, and essentially down the chain contour defined at zero time, is the fundamental assumption of the reptation theory and all its variants.[1,4-8,37,43-45] A schematic picture of the chain motion is presented in Fig. 1.

The scaling with chain length of the various properties can be readily derived.[46] Internally, the chain behaves like a Rouse chain; the only problem is that it must now travel a mean square distance on the order of the contour length, $L \sim n$, in order to relax its conformation. Thus the terminal or longest relaxation time τ_R is obtained from

$$\tau_R \sim L^2/D_{\text{Rouse}} \sim n^3. \tag{6a}$$

However, with respect to the laboratory fixed frame, the molecule has moved appreciably less, on the order of $\langle S^2 \rangle$, if one assumes that the tube obeys Gaussian statistics. Hence the diffusion constant with respect to the laboratory fixed frame is obtained from

$$D \sim \langle S^2 \rangle/\tau_R \sim n^{-2}. \tag{6b}$$

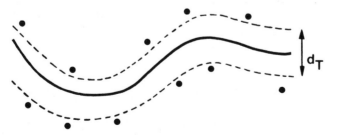

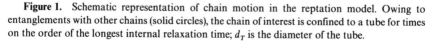

Figure 1. Schematic representation of chain motion in the reptation model. Owing to entanglements with other chains (solid circles), the chain of interest is confined to a tube for times on the order of the longest internal relaxation time; d_T is the diameter of the tube.

Observe that the long chain limit of the self-diffusion coefficient, Eq. (1b), is recovered.

To obtain the dependence on n of the shear viscosity, Doi and Edwards,[6-8] make the further assumption that there is a rubber-like (elastic) response of the melt at short times. Basically, in the short time limit before the chains have a chance to flow, it is impossible to differentiate the behavior of a melt where the entanglements provide the restraining influence, from a rubber where all the chains are covalently cross linked. Unlike the rubber case, the entanglements in a melt are not infinitely long lived, and thus the behavior of a melt is distinct from that of a rubber at longer times. Coupling the assumption of a short time rubbery response with longer time motion down the tube gives

$$\eta \sim \tau_R \sim n^3. \tag{7}$$

Observe that this is not quite the 3.4 power of n, found experimentally for the zero frequency shear viscosity[3,26] (see Eq. 2b); nevertheless, it is close. This has given rise to the conjecture that the $n^{3.4}$ power dependence of η is not the asymptotic behavior, rather it is indicative of a crossover regime.[35,37]

Subsequent work by Graessley provided a means of estimating the magnitude of D and η from reptation theory.[37] The reptation model, typically, overestimates η by about an order of magnitude, but overall it does a rather good job of estimating D.[1,35,37]

Let us compare the behavior of the single-bead autocorrelation function $g(t)$, obtained from a reptation model with the Rouse model.[4,5,46] For distances less than the tube diameter d_T, the chain is unaware that there is a constraining environment, and consequently simple Rouse behavior is recovered. Namely,

$$g(t) \sim t^{1/2} \qquad \text{if } g(t) < d_T^2. \tag{8a}$$

As time further increases, the chain now experiences the restraining effects of the tube and undergoes internal Rouse dynamics in a randomly distributed tube. Hence for times up to τ_{Rouse},

$$g(t) \sim t^{1/4}. \tag{8b}$$

The chain then undergoes free diffusion down the tube, thereby giving

$$g(t) \sim t^{1/2} \qquad \tau_{\text{Rouse}} < t < \tau_R. \tag{8c}$$

Finally, in the free diffusion limit,

$$g(t) \sim t. \tag{8d}$$

Similar considerations indicate that there should be a range of times for which the mean square displacement of the center of mass scales, like $t^{1/2}$ for times less than τ_R but greater than τ_{Rouse}.[20]

While the basic physical picture is quite reasonable for a regular gel, for which the reptation model was originally derived; it is not at all clear the same picture obtains when everything moves on the same time scale. The reality of a spatially fixed tube has been questioned by Fujita and Einaga,[9,10] by Kolinski, Skolnick, and Yaris,[16,17,23,47-49] and Fixman[18,19] for melts and concentrated solutions of linear chains, by Fixman[15] for concentrated solutions of rod-like polymers, and by Baumgartner and Muthukumar[50,51] for chains in a random static medium.

Thus, in the past several years the validity of the original reptation model has been examined.[9-12,15-19,34,47-51] There are two points that need to be clarified. First, is the fundamental assumption that stress relaxation arises from reptation-like motion correct? If so, is there a well-defined tube for motions on the scale of the radius of gyration?.

The examination of what is known about the nature of the dynamics in dense polymer systems is the focus of this chapter, the outline of the remainder of which is as follows. In Section II we present an overview of computer simulation results on multichain dynamics. In Section III we summarize the results of recent analytic theories of polymer dynamics and point out the agreements and disagreements with the simulation results. Section IV concludes the chapter with an overview of the status of the field.

II. COMPUTER SIMULATIONS

A. Dynamic Monte Carlo Results

The problem immediately encountered in an attempt to simulate the long-time dynamics of a dense polymer system is that one must have the ability to study sufficiently long chains for sufficiently long times so that one can verify that the scaling behavior of D and τ_R are reproduced by the simulation. One possible way of simulating systems into the crossover regime is to employ a lattice representation of the polymer melt and perform a dynamic Monte Carlo (MC) simulation.[20,52,53] There are a number of intrinsic advantages as well as disadvantages to this approach; we discuss each in turn.

A dynamic MC simulation consists of the random sampling of configuration space by the following procedure.[52] Starting with an initial configuration of the system, one then chooses a chain at random and then a bead at random. One then randomly displaces, by a set of elemental local moves, the bead of interest. One must be careful that the choice of allowed moves spans the configuration space of the chains, otherwise nonphysical dynamics will result.[22,54-56] In the case of systems interacting solely with a hard-core potential, the move is accepted, provided that two or more beads do not overlap. This method of sampling is known as an asymmetric Metropolis MC scheme. Provided that some additional assumptions are fulfilled, in particular that there must be a path to every state on the system and every step of the process is reversible, then in the limit of a large number of such micromodifications, the system will sample all states with a frequency that is close to their relative Boltzmann probabilities; thus good equilibrium sampling can be obtained.[52,53]

The asymmetric Metropolis MC scheme, when implemented using small-scale local micromodifications of the chain configuration, generates a solution to a stochastic kinetics master equation for the time evolution of the system and, therefore, is able to mimic dynamics.[57] Whether or not the dynamics is physical will depend on the kinds of moves employed. Care must be taken to make the elemental moves as small as possible to avoid the problem of significant time scale distortion. However, there is no guarantee a priori that the chosen moves can mimic physical dynamics. Checks must be performed, such as demonstrating that isolated random coil chains obey Rouse dynamics, that the scaling behavior of D and τ_R with n is recovered, and that the relative mobility exhibited by different chain topologies (e.g., rings and linear chains) tracks experiment.

The advantage of Monte Carlo dynamics (MCD), by setting the fundamental time scale as that required for local conformational modifications, is that it is inherently more efficient than molecular dynamics where the intrinsic time scale is associated with rattling about in local conformational wells. Thus, if one is interested in global relaxation properties of long-chain polymers where presumably such details are unimportant, then MCD is the method of choice.

MCD can, in principle, be performed both on[53,58] and off lattice.[20,21] The advantage of performing MCD on the lattice is twofold. First of all, it allows one to perform the calculations in integer arithmetic, therefore, providing at least an order of magnitude speed up over off-lattice calculations that must be done using floating-point arithmetic. Second, it allows one to rigorously insure that no bond cutting occurs. By use of a lattice occupancy list, this can be done extremely efficiently.[23] Since entangled systems are the object of these studies, it is extremely important that the only way entanglements relax is through physical processes, and not by the nonphysical passing of one chain through the other. The disadvantage is that one is confined to a lattice, and one must show that the results are consistent with off-lattice simulations and that the qualitative results of the simulation are not lattice artifacts. This is always a concern when performing lattice calculations.

In what follows next we present an overview of the results of the diamond[23] and cubic lattice[47,48,59] MCD simulations performed over the past several years by Kolinski, Skolnick, and co-workers for melts of linear and ring polymers; the former simulations are discussed first. As in all simulations of this genre that were previously performed on much smaller systems,[56,60-62] the lattice is enclosed in a periodic box of size $L \times L \times L$. To avoid the problem of a given chain interacting with its periodic image, L is chosen such that it is larger than the root-mean-square end-to-end vector $\langle R^2 \rangle^{1/2}$.[23,48,49] Each polymer chain occupies n lattice sites, and ϕ is the volume fraction of occupied sites. The dynamic properties of homopolymeric (i.e., a melt in which all the chains are equal in length and identical in composition) diamond lattice polymers[23] were examined over a range of volume fraction ϕ from an isolated chain to $\phi = 0.75$, and n ranged up to 216. The cubic lattice polymers were studied at fixed $\phi = 0.5$, for a range of chain lengths $n = 64$ to 800 for the homopolymeric melt.[48]

The first problem one must address, before undertaking the simulation of the dynamics, is the construction of a dense equilibrated melt. The details for preparing such systems has been discussed at length elsewhere.[63] Then one must choose the set of local moves; the importance of this has been discussed above.[54-56]

The local elemental moves that are employed must not only have the ability to diffuse orientations down the chain, but just as in the real system, the ability to introduce locally new orientations into the chain interior as well.[54,55] For the case of diamond and cubic lattices, the set of elemental moves depicted in

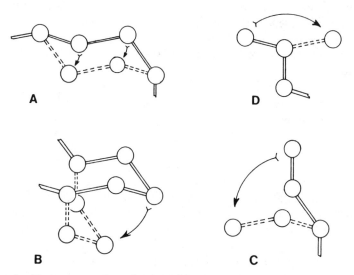

Figure 2. Elementary conformational modifications for diamond lattice polymers. (A) Three-bond kink motion. (B) Four-bond kink motion. (C) One- and (D) Two-bond end motion.

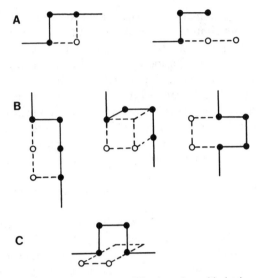

Figure 3. Elementary conformational modifications for cubic lattice polymers. (*A*) Two-bond kink motion and an example of chain-end motion. (*B*) Examples of three-bond kink motions. (*C*) An example of 90° crankshaft motion.

Figs. 2 and 3, respectively, satisfy these criteria.[23,47,48,56,58,62,64] The fundamental time unit is that when each of the beads, on average, is subjected to all the local motions. With this definition of time, the local moves shown in Figs. 2 and 3 give the correct isolated random coil (Rouse) dynamics. As shown below, it is encouraging that in spite of the very different local moves, both the cubic and diamond lattices exhibit the same qualitative behavior, when corrected for differences in local persistence length and lattice coordination number, thereby providing encouragement that the simulation results are physically meaningful.[23,48,49]

1. Center-of-Mass Motion and Longest Internal Relaxation Times

Before even beginning a detailed analysis of the internal chain motion, one must be sure that the scaling behavior of the self-diffusion constant and the terminal relaxation time are consistent with experiment. Figure 4*A* shows, in a log–log plot, the mean square displacement of the center of mass $g_{cm}(t)$ versus time for homopolymeric cubic lattice systems at $\phi = 0.5$.[47] Two distinct time regimes are apparent. For distances such that $g_{cm}(t) < 2\langle S^2 \rangle$, $g_{cm}(t) \sim t^a$ with a decreasing monotonically from 0.91 when $n = 64$ to 0.71 when $n = 800$. Qualitatively identical behavior is found on the diamond lattice[23] as well as in off-lattice simulations.[21] Hence the existence of a t^a regime with $a < 1$ is not a lattice artifact, but is indicative of some kind of constrained dynamics where the center-of-mass motion couples into the internal relaxation processes. This is also consistent with the fact that $2\langle S^2 \rangle$ is the maximum distance over which the internal modes of an individual chain can relax if, in fact, the chains are statistically independent. Clearly then, the behavior of the longer chains is not Rouse-like (see Section I.C).

From the long time slope of the curve shown in Fig. 4A, the diffusion coefficient has been extracted. Fitting the data from $n = 64$ to $n = 216$, one finds a scaling behavior $D \sim n^{-1.52}$. At the time the simulations were done, we lacked the resources to run the $n = 800$ system into the free diffusion limit. Thus a number of extrapolation procedures were employed to extract the power law behavior of D; these are discussed elsewhere.[47] Here we quote the conclusion that the $n = 800$ system is well into the $D \sim n^{-2}$ regime.

The next quantity examined is the longest internal relaxation time τ_R, obtained by standard techniques from the decay of the autocorrelation function of the end-to-end vector. Table I summarizes the scaling of $D \sim (n-1)^{-\alpha}$ and $\tau_R \sim (n-1)^{\beta}$ for the diamond and cubic lattice systems.[23,48] In the case of diamond lattice systems, the deviation in α and β from $\alpha = 2$ and $\beta = 3.4$ arises at lower density from the fact that the chains studied are not long enough to cross over to entangled behavior. Increasing the density at fixed chain length increases the extent of interchain entanglement. For example, the longest chains simulated at $\phi = 0.5$ on the diamond lattice are $n = 216$. This roughly corresponds to a $n = 100$ chain on the cubic lattice, since chains on a diamond lattice are inherently stiffer. Thus longer chain lengths would have to be simulated to observe the asymptotic scaling behavior. Note that the $\phi = 0.75$ chains on the diamond lattice exhibit the requisite scaling.[23] These systems are highly mobile, and no evidence of the slowing down for distances greater than a bond length, indicative of the onset of the glass transition (which does occur at $\phi = 0.92$),[63] is seen. These chains exhibit globally isotropic long-time behavior. Thus the range of densities and chain lengths are appropriate to examine the existence of reptation, since $\alpha = 2.05$ and $\beta = 3.36$, the desired scaling with chain length is found.

Note that at all concentrations the product $D\tau_R$ scales like $n^{1.2}$ for a diamond lattice independent of the concentration[23] and $n^{1.1}$ for the cubic lattice.[47] Based on elementary scaling arguments, $D\tau_R$ should scale like $\langle S^2 \rangle$, which is proportional to $n^{1.0}$.[37] Perhaps this reflects the ± 0.05 uncertainties in the exponents α and β at high densities, or perhaps this is due to the conjectured crossover regime before $D\tau_R \sim n^{1.0}$. The Colby et al.[37] measurements are not inconsistent with this explanation, but they by no means demand it. A third explanation is to take the observation at face value and conclude that $D\tau_R \sim n^{1+\varepsilon}$ with $\varepsilon > 0$. This idea forms the basis of a theory of polymer dynamics introduced by Fixman[18,19] (see Section III.D).

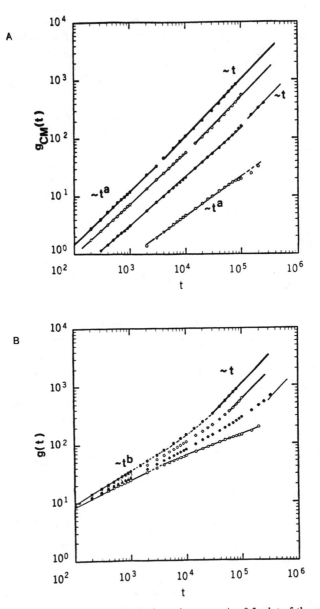

Figure 4. For homopolymeric cubic lattice polymers at $\phi = 0.5$, plot of the mean square displacement of the center of mass, $g_{cm}(t)$ vs. time in (A) and the average single bead displacement $g(t)$ vs. time in (B).

TABLE I
Chain Length Dependence of the Self-Diffusion Coefficient $D \sim (n - 1)^{-\alpha}$ and the Terminal
Relaxation Time, $\tau_R \sim (n - 1)^\beta$ on Cubic and Diamond Lattices

ϕ	a	b
	Cubic Lattice	
0.5	1.52 (± 0.006)	2.63 (± 0.04)
	Diamond Lattice	
Single chain	1.154 (± 0.010)[a]	2.349 (± 0.018)[a]
0.25	1.372 (± 0.021)[a]	2.563 (± 0.061)[a]
0.50	1.567 (± 0.017)[a]	2.677 (± 0.035)[a]
0.75	2.055 (± 0.016)[a]	3.364 (± 0.082)[a]

[a] Standard deviation of the slope obtained from linear least-square fit of log–log plots.

Next, the finer details of the chain dynamics will be examined. In Fig. 4B, the average mean square displacement per bead, $g(t)$, is plotted versus time on a log–log plot for the $\phi = 0.5$ cubic lattice chains.[47] Just as for $g_{cm}(t)$, two distinct time regimes are evident. The first regime, which once again extends up to $2\langle S^2 \rangle$, exhibits a $g(t) \sim t^b$ dependence, where b decreases gradually from the Rouse exponent value of 0.54 when $n = 64$ to a value of 0.48 when $n = 216$. If one stopped increasing the length of the chains at this point, one might reasonably conclude that the dynamics of these chains is entirely Rouse-like. However, this doesn't hold for the $n = 800$ system. Here there is a region where $g(t) \sim t^{0.36}$, once again indicative of the more constrained nature of the dynamics of these chains.

It is not unreasonable to guess that the behavior exhibited by $g(t)$ and $g_{cm}(t)$ is indicative of the crossover to reptation dynamics where one expects a $g(t) \sim t^{1/4}$ regime (see Eq. 8). One might expect that the central beads of the chain would cross over to the $t^{1/4}$ regime first. Thus Figs. 5A and B present log–log plots of the average mean-square displacement of the central five beads in the chain $g_5(t)$ versus t for $n = 216$ and $n = 800$ chains, respectively. There is clear evidence for a $t^{1/4}$ regime. The fundamental question still remains; namely, are these chains reptating? To get slightly ahead of the story, as shown in the next section, a detailed microscopic analysis of the microscopic motions of these chains indicates that the character of the chain motion is entirely different.

In fact, the existence of a $t^{1/4}$ regime is indicative of some sort of constrained interchain dynamics and is *not a unique signature of reptation*. This statement is further substantiated by a recent simulation of Milik et al.[65] on a model of a membrane, confined to a diamond lattice. The head of each molecule is constrained to bob up and down, no more than one lattice unit in the z

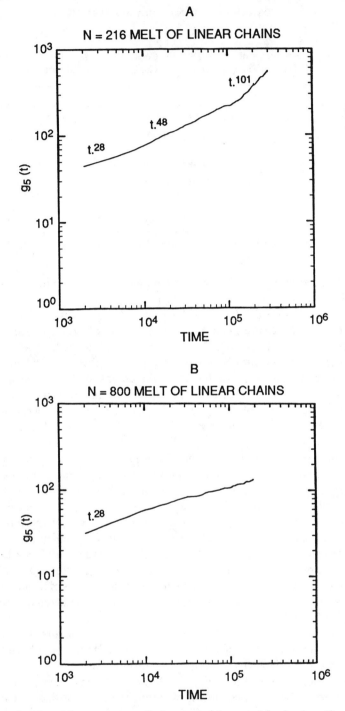

Figure 5. Plot of the mean square displacement of the central five beads, $g_5(t)$ vs. time for $n = 216$ (A) and $n = 800$ (B) in $\phi = 0.5$, homopolymeric, cubic lattice melts.

direction, but the chains are free to move in the other two dimensions. Given the constraints on the dynamics, these chains cannot possibly reptate (one tail is more or less nailed down in one dimension). There is the standard $t^{1/2}$ regime; this is followed by a $t^{1/4}$ regime at distances corresponding to the mean spacing between chains. Hence the existence of a $t^{1/4}$ regime in the single bead autocorrelation function does not prove the existence of reptation, since systems which cannot possibly reptate exhibit this behavior as well.

2. Examination of the Primitive Path Dynamics

In their classic papers developing many of the essential ideas of the reptation model, Doi and Edwards[5-8] invoke the idea of a primitive chain path.[66] The primitive path entails the replacement of the actual chain by an equivalent in which all the local fluctuations in the chain contour which are irrelevant to the long distance motion are averaged out. Conceptually, this is much like taking the original chain contour, reeling in the slack, and examining the resulting path.

To examine the trajectories of the chains in the simulation, we have constructed an equivalent chain and followed its motion as a function of time.[23,47,48] The outline of the procedure is as follows: Each bead in the original chain is replaced by a point on the equivalent chain which is the center of mass of a subchain composed of n_b beads. This replaces the actual chain contour by a smooth path of partially overlapping subchains, which should be a good approximation to the primitive path of Doi and Edwards if n_b is on the order of the number of monomers between entanglements. (Actually, the results described below are quite insensitive to n_b.) The equivalent path is generated as a function of time. At every time the equivalent path is projected onto the original path defined at zero time. The displacement down the original path corresponds to the reptation component. What remains is the nonreptation component of the dynamics, which should be small if reptation is dominant.

To quantify the measurement of reptation, the mean-square displacements down the original primitive path, $g_{\parallel}(t)$, and perpendicular to the original primitive path, $g_{\perp}(t)$, are calculated. If the chain reptates, it is straightforward to show that the maximum value of $g_{\perp}(t)$ equals one-half the mean-square tube radius for times less than the tube renewal time.[23] Thus the ratio $g_{\perp}(t)/g_{\parallel}(t)$ should monotonically decrease with increasing time. If, however, the motion is isotropic and liquid-like with little if any memory of a tube defined at zero time, then $g_{\perp}(t)/g_{\parallel}(t)$ should monotonically increase. Thus, examining this ratio is a nonbiased way of directly addressing the question of whether or not a given system of chains reptates in a fixed tube. It is interesting to point out that reptation theory assumes that a kind of glass transition is occurring in the melt. That is, the motion of the chain perpendicular to the original path is essentially frozen out due to the existence of entanglements.

Since the simulations involve MCD on a lattice, it is important to establish that somehow reptation is not artificially suppressed by the choice of elemental moves and to verify that the ratio $g_\perp(t)/g_\parallel(t)$ does indeed indicate reptation when in fact it is present. It is well established that if a single chain is confined to a fixed mesh and if it is sufficiently long, the chains will reptate.[21,43,44]

To validate the simulation methodology, we examined the dynamics of a single chain in a partially frozen environment.[23] Basically, what one does is take the original $n = 216$ diamond lattice polymer $\phi = 0.5$ melt, and freezes all but one test chain in place. However, if all the matrix chains are completely frozen, since the tube is not porous, the test chain is trapped. Thus a partially frozen environment was employed, where every eighteenth bead in the matrix chains is frozen in place. This allows for local dynamics that are quite close to the original melt, but where all the chains but the test chain are constrained from moving appreciable distances.

Examination of the primitive path revealed that the chains reptate and the ratio $g_\perp(t)/g_\parallel(t)$ versus time monotonically decreases, as expected.[23] The signature of reptation is recovered, and one finds the expected presence of reptation. Hence the MCD moves that are used do not somehow artificially suppress reptation. We do note, however, that there is substantial tube leakage,[1] with loops running up and down cul de sacs. A further interesting point, that is not surprising, is that the chain in the partially frozen environment moves substantially slower than when all the chains are free to move.

In Fig. 6 the ratio $g_\perp(t)/g_\parallel(t)$ versus t is shown for the cubic lattice melt,[47] where everything moves, for chains with $n = 216$ and $n = 800$. Here n_b has been set equal to 17 as well as 101, and no qualitative difference is found.

The qualitative features exhibited in Fig. 6 are identical to those seen for chains on the diamond lattice.[23] At short times transverse motion of the chains is preferred. This results from the nature of cooperative motions at high density, whose origin is as follows.[63] Imagine a chain has undergone a conformational rearrangement. The probability of the chain undergoing correlated motion is the product of two quantities: (1) the intrinsic probability that the chain is in a conformation in which a conformational change is possible and (2) the probability that there are unoccupied sites, into which the chain can jump. For both cross- and down-chain motion, the intrinsic probabilities are identical. However, for cross-chain motion, given that the chain has already undergone a jump, there is now an unoccupied volume that the neighboring chain can jump into. The conditional probability that the neighbor can undergo the jump is unity. In the case of down-chain motion, this probability is to lowest order proportional (on a lattice) to $(1 - \phi)$ raised to the power of the number of sites involved in the motional unit. Therefore, with an increase in density, one would expect that cross-chain motion is dominant at short

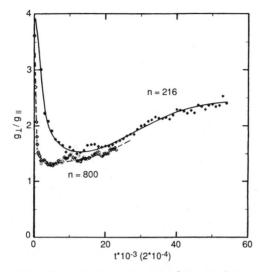

Figure 6. Plot of the ratio $g_\perp(t)/g_\parallel(t)$ vs. time $\times 10^{-3}$ (2×10^{-4}) for $n = 216$ (800) in the upper (lower) curve. Both are at a density of $\phi = 0.5$ on a cubic lattice.

times, as is observed. This qualitative conclusion is independent of whether or not the chains are on a lattice.

Subsequent to the short time preference for transverse motion, there is a period when down-chain motion becomes more important. This corresponds to distances on the order of the excluded volume decay length. For these distances, the chain starts to feel the effect of the environmental constraints and has slowed down. There is a certain incubation period before the collective motion of the chains that gives rise to the lateral motion takes over. Finally, at longer times the reptation component becomes less important and the lateral component grows. In fact, it becomes increasing difficult to follow the original primitive path and project on it. This is true, in spite of the fact that according to reptation theory this is precisely the distance and time regime for which reptation dynamics should be very well defined.[1,4-8,37] Note that we have only examined the primitive path for the middle third of the chain, because any chain, whether reptating or not, undergoes substantial fluctuations of the ends.[17,23]

An interesting point observed on comparison of Fig. 6 with Fig. 5A and B is that the minimum in $g_\perp(t)/g_\parallel(t)$ versus t occurs just as the chains are crossing out of the $t^{1/4}$ regime in the $g_5(t)$ versus t plot. Thus if one were to merely look at $g_\perp(t)/g_\parallel(t)$ for times just up to the end of the $t^{1/4}$ regime, but for motion over distances still small relative to the radius of gyration, one would incorrectly conclude that reptation is quite important. It is not until the chain goes further into the second $t^{1/2}$ regime that reptation becomes a minor component of the dynamics. Thus a recent off-lattice molecular dynamics simulation,[24] which claims to see reptation, is in fact, inconclusive because only times up to the end of the $t^{1/4}$ regime are sampled. The simulation times are too short to demonstrate whether the chains are reptating or not. As discussed above, the existence of a $t^{1/4}$ regime in $g(t)$ versus t plots is insufficient to prove the existence of reptation. *Other modes of cooperative dynamics give this result as well.*

A more pictorial illustration of the character of chain motion is shown in Fig. 7A–C, where the trajectory of one of the $n = 800$ chains confined to a cubic lattice is presented. The thin line corresponds to the conformation at the initial time, and the thicker line shows the conformation at a time t, later. The triangle indicates the position of one of the chain ends. For ease of visualization, n_b was set equal to 101. Consistent with the ratios $g_\perp(t)/g_\parallel(t)$ versus t, substantial lateral fluctuations are evident. One is forced to conclude that these chains, at least, do not know that they are confined to a static tube.

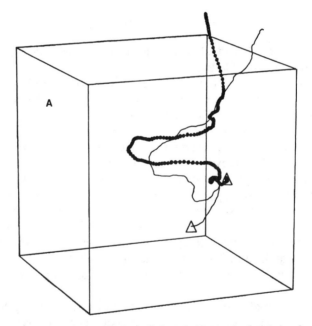

Figure 7. $(A–C)$ Snapshot projections of the primitive path of a chain of $n = 800$ in the $\phi = 0.5$ cubic lattice melt. The thin line corresponds to the initial conformation and the triangle labels one of the ends. The displacement after 6×10^4 steps, 1.2×10^5 steps, and 2.0×10^5 steps is shown in $A–C$ respectively.

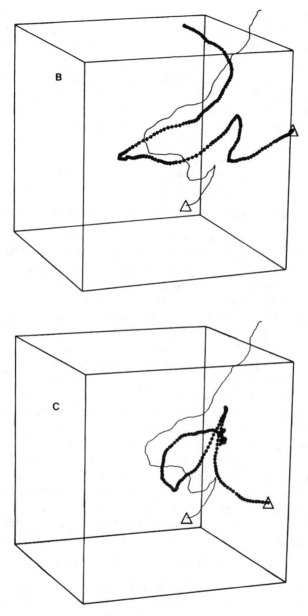

Figure 7. (*Continued*)

The question still remains as to whether or not real chains reptate. Certainly, these simulations strongly argue against the existence of a fixed tube, but it could be argued that these chains are in the crossover regime from Rouse to reptation dynamics. In this regime, one might still expect substantial tube fluctuations; in other words, we are observing a regime where the chains do not *yet* reptate. Since the $n = 800$ chains are about at the limit of our computational capabilities, we cannot rule out a crossover to fixed-tube/reptation behavior at increased chain length. However, because the present simulations reproduce the experimental scaling of D and τ_R, it would have to be a crossover from a regime where classical reptation is not dominant to a regime where reptation dominates which is invisible to experiment, at least via the standard techniques that have been employed.

B. Probe Polymer in Matrices of Different Molecular Weight

Another important observation is that when a test or probe polymer is dissolved in a matrix of polymers of identical chemical composition but increasing molecular weight, the diffusion constant of the probe, D_p, becomes independent of matrix molecular weight.[28,29,67] To examine whether the present model system could reproduce this behavior, the dynamics of a probe chain composed of $n_p = 100$ segments in matrices from $n_m = 50$ to 800 segments was explored in a cubic lattice system having $\phi = 0.5$.[48] Over this range of matrix molecular weights, the diffusion constant of the probe decreases by approximately 30%; this is consistent with the decrease in D_p observed in real experiments.[27] While a chain of $n_p = 100$ is not sufficiently long to have crossed over into the n^{-2} regime of the diffusion constant, the probe in the $n_m = 800$ system had a diffusion constant that is two orders of magnitude larger than the matrix in which it was dissolved.

The primitive path analysis clearly showed that the motion of the chain is not confined to a tube and that the motion is not reptation-like. Otherwise stated, the local fluctuations in the topological constraints imposed by the matrix are sufficiently large even for the $n_m = 800$ case to allow for essentially isotropic, but somewhat slowed-down, motion of the probe chains in the melt. Thus the MCD is once again in qualitative accord with experiment, and yet reptation within a confining tube is not found.

C. MCD Simulation of Melts of Rings

More recently, Sikorski, Kolinski, Skolnick, and Yaris[59,68] undertook a series of simulations designed to examine the nature of the dynamics of a melt of uncatenated rings; two distinct physical cases were examined. The first, and simplest, involved a cubic lattice melt of unknotted cubic lattice rings, packed at a volume fraction $\phi = 0.5$; a range of chain lengths from $n = 100$ to 1536 was studied. The second case involved the dynamics of a melt of rings

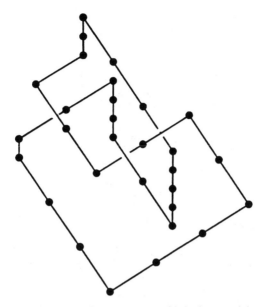

Figure 8.   Conformation of an $n = 30$ ring on a cubic lattice containing one self-knot.

containing one self knot, such as is shown in Fig. 8. The motivation of this simulation, suggested to us by McKenna, is as follows. Owing to the nature of the synthesis conditions, it is possible that real ring polymers might contain self knots, and the question arises as to whether such knots will drastically change the dynamic properties of a melt of rings.[39]

1.   Growth of Melts of Rings

Just as in the linear case, one has to start with an equilibrated melt of rings before one can even begin to perform the dynamics. Recently, Pakula and Geyler developed a very efficient method for generating such a system on a fully occupied lattice.[69-71] This novel algorithm will be elaborated on in detail in Section II.E, where their simulation technique and results are more fully explored. Here we describe an algorithm for generating systems, in principle of arbitrary polydispersity, which should be quite efficient, provided that the density of the system is not too large.

As schematically depicted in Fig. 9 for the case of unknotted rings, one starts with a set of the smallest size rings, each containing four beads. One chain is shown for the sake of clarity; actually, N chains are simultaneously subjected to the growth/equilibration algorithm. These are then randomly grown and modified, just as is done for linear chains. However, unlike the

A

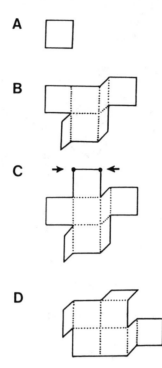

B

C

D

Figure 9.   Schematic representation of the ring growth algorithm. One ring is shown for clarity. In reality all rings are simultaneously grown and equilibrated.

linear case,[63] reptation dynamics is useless (there are no free ends to randomly cut and paste). Fortunately, since rings are characterized by smaller chain dimensions and for isolated chains, at least, smaller relaxation times, this doesn't pose too severe a problem. That is, the systems can be run for sufficiently long times to be reasonably sure that equilibrated systems have been prepared. The rings undergo the same set of internal modifications as the linear chains shown in Fig. 3, with the exception that end moves are not performed. The process of growth/equilibration stops when all the rings are of size n. In principle, polydisperse melts of rings could be grown.

The advantage of this particular ring preparation procedure is that owing to the set of elementary jumps of MCD and the ring growth mechanism the rings cannot be catenated. Depending on the degree of polymerization and density, the rings can be rather highly entangled. Provided that the equilibration period between intervening growth steps is sufficiently long, the resulting system should be close to an equilibrated system. Following system preparation, the system is allowed to run for several relaxation times before sampling for the dynamic properties begins.

The only difference between growing rings that have no knots and rings with a single knot is in the starting conformation. Initially, rings having a single knot are placed in the box, and then they are subjected to the growth/ equilibration process. Again, because of the bond pair insertion constraint that does not violate excluded volume, a monodisperse collection of rings, having one self-knot, can be generated as well.

2. Equilibrium Properties

To date, only two simulations of the equilibrium properties of a melt of rings have been undertaken, one by Pakula and Geyler[71] for rings on a fully filled cubic lattice, up to $n = 512$, and the present simulation for rings at $\phi = 0.5$, for rings up to $n = 1536$.[59,68] Qualitatively, identical behavior is observed. Before presenting the simulation results, it is appropriate to compare the expected scaling relationships of the conformational properties with the corresponding linear chains, if the statistics are Gaussian.[42] The mean square radius of gyration of the corresponding ring and linear chain of identical n are related by

$$\langle S^2 \rangle_{\text{ring}} = \langle S^2 \rangle_{\text{linear}}/2 \sim n^{1.0}, \tag{9a}$$

and the mean square diameter, which is the mean square distance between beads 1 and $n/2$, $\langle d^2 \rangle_{\text{ring}}$, is related to $\langle R^2 \rangle_{\text{linear}}$ by

$$\langle d^2 \rangle_{\text{ring}} = \langle R^2 \rangle_{\text{linear}}/4 \sim n^{1.0}. \tag{9b}$$

Combining Eqs. 9a and 9b,

$$\langle S^2 \rangle_{\text{ring}}/\langle d^2 \rangle_{\text{ring}} = \tfrac{1}{3}. \tag{9c}$$

These results are to be compared with the Pakula and Geyler[71] simulation result that

$$\langle S^2 \rangle_{\text{ring}} \sim n^{0.90}, \tag{10a}$$

and

$$\langle R^2 \rangle_{\text{linear}} \sim n^{1.006}, \tag{10b}$$

the latter holding for chains up to $n = 512$.

The Sikorski et al. result[59,68] is

$$\langle S^2 \rangle_{\text{ring}} \sim n^{0.84}, \tag{11a}$$

and

TABLE II
Equilibrium Statistics of Melts of Rings at $\phi = 0.5$

n	N	$\langle d^2 \rangle$	$\langle S^2 \rangle$	$\langle S^2 \rangle / \langle d^2 \rangle$	$\langle d^4 \rangle / \langle d^2 \rangle^2$	$\langle S^4 \rangle / \langle S^2 \rangle^2$	$\langle d^2 \rangle_{ring} / \langle R^2 \rangle \, \text{lin}^b$	$\langle S^2 \rangle_{ring} / \langle S^2 \rangle_{lin}$
100	20	46.9 ± 0.04	15.15 ± 0.07	0.323	1.540 ± 0.008	1.070 ± 0.002	0.250	0.492
100^a	20	28.0 ± 0.8	11.60 + 0.02	0.414	1.666 ± 0.02	1.041 ± 0.001	0.149	0.377
216	24	91.8 ± 0.3	30.0 ± 0.3	0.336	1.587 ± 0.017	1.081 ± 0.003	0.222	0.458
392	28	154.6 ± 7.9	52.8 ± 1.4	0.342	1.597 ± 0.026	1.083 ± 0.006	0.215	0.424
392^a	28	132.5 ± 10.0	45.4 ± 1.7	0.347	1.664 ± 0.020	1.084 ± 0.009	0.184	0.369
800	40	278 ± 19	95.2 ± 3.0	0.341	1.625 ± 0.077	1.091 ± 0.020	0.195	0.368
1536	40	469 ± 50	168.4 ± 5.6	0.359	1.657 ± 0.101	1.109 ± 0.020	0.169	0.333

[a] Melt of rings each containing one self knot.
[b] lin is the value for the linear system.

$$\langle d^2 \rangle_{\text{ring}} \sim n^{0.87}, \tag{11b}$$

which should be compared to the linear melt values[47] of

$$\langle R^2 \rangle_{\text{linear}} \sim n^{0.99}, \tag{12a}$$

and

$$\langle S^2 \rangle_{\text{linear}} \sim n^{1.02}. \tag{12b}$$

Thus the $\phi = 0.5$ result is in excellent agreement with the $\phi = 1.0$ result of Pakula and Geyler.[71] With respect to the equilibrium properties at least, the $\phi = 0.5$ system is essentially identical in scaling behavior to the fully occupied lattice system.

In Table II, the Sikorski et al. equilibrium results[59,68] are summarized for both the unknotted system and the system having one self knot. The latter systems, not unexpectedly, have smaller dimensions than the unknotted rings, but the relative difference between the two cases decreases with increasing chain length.

The simulation results of both groups are in reasonable agreement with the Flory-like mean field treatment of Cates and Deutsch,[72] which gives

$$\langle d^2 \rangle_{\text{rings}} \sim n^{4/5} \tag{13}$$

in three dimensions. The crux of their argument is as follows. If excluded volume interactions are fully screened, the fact that catenated ring conformations are prohibited exerts a topological constraint on the system, and the more extended the conformation is, the greater is the topological constraint. Thus rings in a melt should tend to collapse. Opposing this is the Gaussian entropic force. Equation (13) results from minimizing the free energy due to these two competing effects, with the result that rings are less expanded than would be predicted from Gaussian statistics alone.

One of the more interesting and unexplained relationships of Table II is that while the $\langle S^2 \rangle_{\text{rings}}$ and $\langle d^2 \rangle_{\text{rings}}$ are distinctly non-Gaussian in behavior, their ratio is close to the Gaussian value of $1/3$.

3. Dynamic Properties of Unknotted Rings

The procedure for extracting the self-diffusion coefficient and the longest relaxation time τ_d (which corresponds to the relaxation of the diameter autocorrelation function) is discussed elsewhere.[59,68] Values of D, τ_d, and a number of other dynamic quantities for rings of $n = 100, 216, 392, 800,$ and 1536 are shown in Table III. Given the computational resources, we cannot carry out

TABLE III
Self-Diffusion Constant and Terminal Relaxation Times for $\phi = 0.5$ Melt of Rings

M	D	τ_d	$D\tau_d/\langle d^2 \rangle$	$n/n_e{}^b$	a^c	b^d
100	1.14×10^{-3}	1.15×10^3	2.80×10^{-2}	0.20	0.96	0.61
100^a	9.15×10^{-4}	1.17×10^3	3.82×10^{-2}		0.95	0.56
216	4.35×10^{-4}	7.50×10^3	3.55×10^{-2}	0.42	0.88	0.52
392	1.88×10^{-4}	3.02×10^4	3.67×10^{-2}	0.77	0.87	0.45
392^a	2.85×10^{-4}	2.43×10^4	5.23×10^{-2}		0.90	0.46
800		1.43×10^5		1.57	0.85	0.46
1536		5.89×10^5		3.01	0.80	0.41

[a] Melt of rings each containing one self-knot.
[b] Estimated from Eq. (15a).
[c] Exponent of $g_{cm}(t) \sim t^a$ for distances less than $2\langle S^2 \rangle$.
[d] Exponent of $g(t) \sim t^b$ for distances less than $2\langle S^2 \rangle$.

the $n = 1536$ simulation into the free diffusion limit. The diffusion constant scales with n like

$$D \sim n^{-1.42},$$ (14a)

and

$$\tau_d \sim n^{2.32}.$$ (14b)

This scaling behavior suggests that the rings are in the crossover regime and are not as entangled as the corresponding linear system.

Another means of estimating the number of entanglements follows from an analytically derived expression for the diffusion constant of a chain in the melt[16] (see Section III.B below)

$$D(n) = \frac{d_0}{n\left(1 + \dfrac{n}{n_e}\right)},$$ (15a)

where n_e is the mean number of monomers between entanglements. Fitting this expression to data in Table III gives n_e (rings) = 515, and employing the analogous expression to fit the linear data[47,48] gives

$$n_e(\text{rings})/n_e(\text{linear}) = 3.9.$$ (15b)

In other words, the melt of rings is less entangled and therefore, more mobile than the corresponding melt of linear chains. This ratio compares quite favorably to the ratio of approximately 5 obtained for polybutadiene;[40] other ratios reported for polystyrene are about two.[39] Hence the simulations once again reproduce the experimental trends.

Unfortunately, since rings have a larger n_e than the corresponding linear chains this means that it is presently impractical to simulate rings of a comparable (albeit relatively small) degree of entanglement as the corresponding linear chains on available computers. The decrease in the number of entanglements is partially due to the more compact dimensions of rings as

A

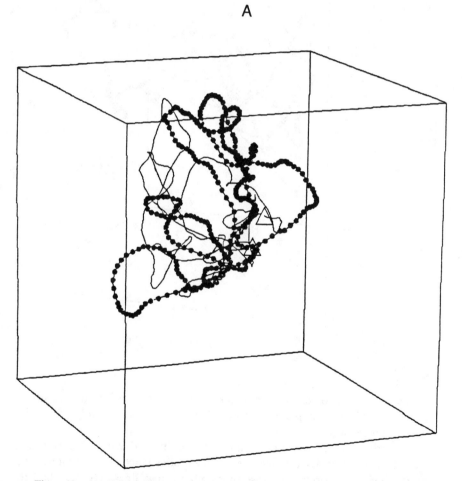

Figure 10. $(A-C)$ Snapshots of the primitive path of an $n = 1536$ ring in the $\phi = 0.5$ cubic lattice melt. The thin line denotes the initial conformation, and the triangle labels beads 1 and 1536. The displacement after 3.04×10^5, 8×10^5, and 1.008×10^6 steps is shown in $A-C$ respectively.

B

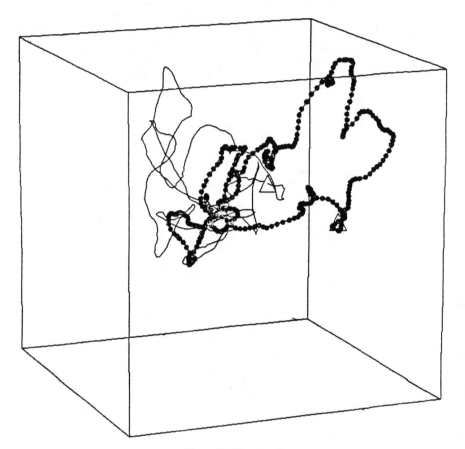

Figure 10. *(Continued)*

compared to the corresponding linear chains. Another effect may be due to the nature of the entanglements themselves, a point that we address further in the next section.

For advocates of reptation, rings being entirely devoid of ends pose a particular problem. Klein[38] has proposed a model which asserts that the only rings which are mobile are those that collapse to a linear chain of half the contour length. If this expectation is true, one would expect to find almost all rings immobile and an exponentially small subpopulation that is highly

C

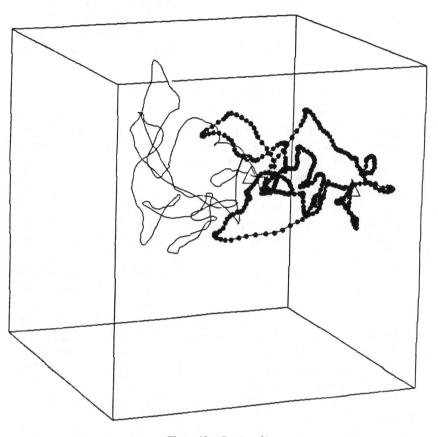

Figure 10. *(Continued)*

mobile. Thus we examined the population of mobile rings to see if such a bimodal population exists; none was found.

We next present the primitive path as a function of time for the $n = 1536$ system in Fig. 10A–C. The triangle labels beads 1 and 1536. The thin line indicates the initial primitive path obtained at zero time, and the solid line labels the path at a time t later. Nothing striking or reptation-like is seen— rather rings seem to move much like amoebas.

In all fairness to the advocates of reptation, these systems are weakly entangled; nevertheless, if reptation were the dominant mechanism of melt dynamics, one would have expected the onset of reptation for rings to occur at smaller—and not larger—degrees of entanglement than in the corresponding linear chain case. However, because these systems are early in the crossover regime, a transition to much slower dynamics cannot be ruled out—although there is no hint of any such transition in the simulations performed thus far.

4. Properties of Self-Knotted Rings

We next turn to the behavior of the dynamics of a melt of rings, each containing one self knot. Care has to be taken when comparing the dynamics of self-knotted rings with unknotted rings because of an artifact of lattice dynamics. Consider a ring containing one self knot, the minimum ring size that can fit onto a cubic lattice contains 22 beads and is left absolutely immobile when subjected to the internal moves of Fig. 3. This is a close-packed object that can only undergo rigid body translations and/or rotations, and these have not been incorporated into the present MCD algorithm. Thus, one should expect two competing effects as n increases. First of all, internal conformational transitions become possible, and thus D increases from zero. This effect is as indicated above, an artifact of the MCD algorithm that is employed. Second, as polymeric effects take over, D should decrease with increasing molecular weight. Finally, D for isolated chains without knots should always be larger than D with knots (since the conformations of the latter are more compact). When comparing the results of the dynamics of knotted versus unknotted rings in the melt, one wants to be sure that artifacts are eliminated, and we therefore examine the dynamics of isolated rings first to be sure such artifacts are not present.

In the solid (open) circles of Fig. 11, results of D versus n for isolated rings containing one (no) self knot are shown on a log–log plot. Beyond $n = 100$, and certainly for n greater than or equal to 216, D for the knotted system is well defined and monotonically decreases with increasing n.

Fitting the log–log plot for $n = 216, 800,$ and 1536, we find for the knotted system that

$$D \sim n^{-1.3}, \tag{16a}$$

and for the terminal relaxation fit over the range of $n \geq 100$ that

$$\tau_d \sim n^{2.2}. \tag{16b}$$

By way of comparison, the equilibrium quantities are

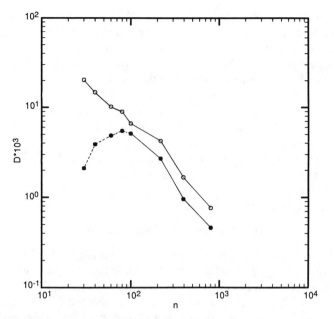

Figure 11. Log–log plot of the self-diffusion coefficient of isolated ring polymers confined to a cubic lattice vs. n. The solid (open) circles are for rings having one (no) self-knots.

$$\langle S^2 \rangle = 0.0411(n - 1)^{1.28} \tag{16c}$$

and

$$\langle d^2 \rangle = 0.1090(n - 1)^{1.28}. \tag{16d}$$

The corresponding quantities for the isolated unknotted rings, are for a fit over the range 30–1536,

$$D \sim n^{-0.96}, \tag{17a}$$

$$\tau_d \sim n^{2.1}, \tag{17b}$$

$$\langle S^2 \rangle = 0.1079(n - 1)^{1.16}, \tag{17c}$$

and

$$\langle d^2 \rangle = 0.3266(n - 1)^{1.19}. \tag{17d}$$

Based on these results we can probably safely compare the dynamics of melts of self-knotted rings with those of unknotted rings for $n = 216$ or greater. Table III also presents a summary of the dynamic properties of melts of self-knotted rings of $n = 100$ and 392. Comparing the knotted with the unknotted rings, it is immediately apparent that within the error of the simulation the dynamics of a melt of self-knotted and unknotted rings are essentially identical. Probably because self-knotted rings are smaller (having $\langle S^2 \rangle$ of about 133 vs. $\langle S^2 \rangle$ of 155, for $n = 392$) than unknotted chains, the self-knotted rings are less entangled and, therefore, slightly more mobile (this competes with the intrinsically lower mobility of isolated self-knotted chains; the latter effect should be more important at smaller n, as is observed). Nevertheless, the effect is minor and we conclude that self knotting will have a marginal effect, at best, on the melt dynamics.

D. The Origin of Entanglements

Whatever the physical origin of the interchain entanglements, to exert an influence on the long distance motion, they must live for times on the order of the terminal relaxation time or perhaps longer. Otherwise, they can be subsumed into a molecular weight independent, monomeric friction coefficient and therefore, they wouldn't change the scaling with molecular weight of the transport coefficients. Based on the results for the linear chain simulations which in many, but not all, respects behave like slowed-down, Rouse chains,[23,47,48] one might conjecture that the slowdown in behavior results from dynamic entanglement contacts. That is, one chain drags another chain for times on the order of the terminal relaxation time. Eventually, of course, these entanglements should disengage. This is an old idea in the literature, which goes back to Bueche,[2,73] variants of which have been proposed by Fujita and Einaga[9,10], Ngai et al.,[14,74] Kolinski et al.,[16,17,23,47,48] and Fixman.[18,19] Thus we next examine what the simulations have to say about this conjecture.

1. Bead Distribution Profiles

In Fig. 12 the time dependence of the mean-square displacement of the ith bead $g_i(t)$ is plotted as a function of the position i, along the chain for $n = 216$ linear chains, packed at $\phi = 0.5$ diamond lattice melt.[23] In the curves denoted by a–d, the time equals 3×10^4, 6.9×10^5, 1.35×10^6, and 2.1×10^6 time steps ($\tau_R = 5.2 \times 10^5$). The smooth curves are generated by using the Rouse model and an apparent diffusion coefficient defined as $g_{cm}(t)/6t$. The Rouse model is seen to overestimate the mobility of the chain interior and underestimate the mobility of the ends. It should be pointed out here that at infinite time in the absence of excluded volume, the bead distribution profile will be parabolic, *independent of the particular model of the dynamics*.

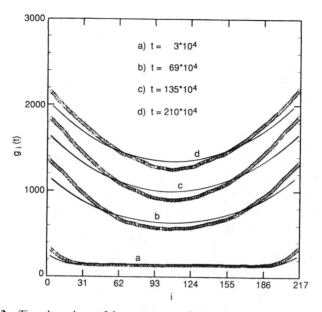

Figure 12. Time dependence of the mean-square displacement of the ith bead as a function of position along the chain for $n = 216$ linear chains confined to a diamond lattice at $\phi = 0.5$.

2. Nature of the Contacts between Chains

To examine the time evolution of the contacts between chains, the following algorithm is employed.[16] (1) Each chain is replaced by a series of nonoverlapping blobs, each having n_b monomers. (2) All pairs of blobs belonging to different chains, whose centers of mass lie a distance less than $r_{min} = 5$, are identified. (The length of a bond equals unity.) (3) The number of such contacts is counted. (4) The fraction of contacts, $n_c(t)$, that survive up to a time t later, given that the blobs were in contact at zero time, is determined.

For the $n = 216$, $\phi = 0.5$ cubic lattice chains, $n_c(t)$ is found to be decomposable into a sum of three exponentials.[16] While this decomposition is of course not unique, the results are highly suggestive. The majority of the contacts (64%) decay within 1% of τ_R, 91% decay within 9% of τ_R, and the remaining contacts decay on the order of τ_R. This translates into one long-lived contact every 133 beads. If Eq. (15a) is used to extract the number of monomers between entanglements, one finds a value of 125. The mean lifetime of these contacts is consistent with the idea that long-lived contacts between polymers slow down the motion of the chains at long times. Most local contacts are short lived, and apart from modifying the local monomeric friction constants, they exert no effect on the long-time dynamics.

The first conclusion that emerges from the simulation is that long-lived dynamic entanglement contacts occur on a distance scale which is an order of magnitude larger than the static excluded volume screening length. In real polymer melts, the excluded volume screening length is on the order of a monomer unit;[25] whereas, based on estimates from the plateau modulus and the crossover regime of the shear viscosity, the mean number of monomers between entanglements is on the order of 100 monomers or so.[3, 26] Moreover, by examining the displacement of the contacts, we have established that all entanglements are moving with respect to the laboratory fixed frame and that there is no static cage. If these model chains were reptating, in a fixed tube no single contact should live on the order of the terminal relaxation time.

What then might dynamic entanglements be? Suppose that at zero time a pair of chains are in a configuration where one chain loops around another chain. This may be a necessary, but not sufficient, condition for the formation of an entanglement. Subsequently, the pair of chains must move in a direction that causes the entanglement to be long lived; that is, one chain drags another chain for times on the order of the terminal relaxation time. Whether or not the disengagement must occur by reptation is not yet established, nor is it clear what sort of configurations cause the dynamic entanglements. The analysis of these kinds of questions is now underway; however, at best, the results must be viewed as tentative. Even in the $n = 800$ linear chains, there are at most 6–7 entanglements per pair of chains, and their statistics are likely to be poor.

E. Cooperative Relaxation Dynamics

In a series of papers Pakula and Geyler,[69-71] have developed a method of simulating chains on a fully packed lattice, using a cooperative exchange mechanism that works as follows. (1) A kink on a given chain is chosen, say chain A. This kink is sliced out of the chain and inserted into a section of chain B that is locally parallel to the top section of the original kink in chain A. Observe that this decreases the length of chain A, and temporarily increases the length of chain B. (2). The algorithm then searches for a similar interchange between chain B and another chain C. (3) The loop replacement procedure is continued until a loop is interchanged back with the original chain A. Thus, at the end of a loop migration procedure, all the chains have been restored to their original contour length. While there is no doubt that this is a highly efficient algorithm for equilibrium sampling, it is not clear how reliable the resulting dynamics are, especially since there is no limitation on the lengths of the mobile loops. They choose 10^6 searching steps as the fundamental time unit; this particular choice, at short time intervals, produces equal mean square displacements of the monomers that are independent of chain length. A possible problem encountered if this algorithm is employed to obtain dynamic sampling is that in a single time unit, interchanges over short and long wavelength distance scales are occurring, and it is not at all clear how (or even if) the time scale is distorted.

Applying this cooperative motion model to a melt of linear chains confined to a cubic lattice, chains up to $n = 512$ were simulated.[70] For longer chains, the self-diffusion coefficient is found to have a $D \sim n^{-2}$ regime, and the end-to-end vector autocorrelation functions are found fit a stretched exponential form. The single-bead autocorrelation function is found to be close to that predicted by the Rouse model. Snapshots of individual chain configurations indicate globally isotropic dynamics with no evidence of reptation.

They next applied their algorithm to examine the dynamics of a melt of rings;[71] the equilibrium results have been summarized in Section II.C.2. They find that rings have a *smaller* self-diffusion constant than the corresponding linear chains, and while this difference decreases with increasing n, it disagrees with the viscoelastic experiments which indicate that melts of rings are more mobile than the corresponding linear chains. The origin of the difference between their results and the local MCD simulations of Section II.C.2 is not clear. One thing the linear simulations point out, however, is that a D proportional to n^{-2} appears to be a ubiquitous result that cannot be invoked as proof of chain reptation. This point is addressed further in Section III.B.1.

F. Dynamics of Chains in Random Media

To examine the fundamental validity of the reptation model Muthukumar and Baumgartner[50,51] have returned to the original system on which the reptation idea is based and have performed off lattice, MC simulations on a single chain diffusing through a random medium. The mean-square displacement of the center of mass of the chain is observed to exhibit three time regimes. At long and short times $g_{cm}(t)$ is linear in time. The duration of the intermediate time regime increases as either the chain length or the volume fraction of the solid phase increases. The short-time apparent diffusion constants are Rouse-like with $D_{app} \sim n^{-1}$. The diffusion constants extracted from the long-time behavior of $g_{cm}(t)$ do not obey the n^{-2} scaling predicted on the basis of the reptation model. Rather, the results indicate the presence of entropic barriers, arising from the necessity of the chain squeezing from one domain to the other through bottlenecks. Thus, even in the case of a static random medium, it does not automatically follow that a chain must reptate. While it is true that as the contour length of the chain increases to the point that a given chain may be in multiple domains, then reptation-like motion will dominate, it is not necessarily true that classical reptation theory should apply. The latter situation appears to hold in the problem of DNA gel electrophoresis and has been analyzed by Levene and Zimm.[75] A further discussion of the theoretical problems associated with DNA gel electrophoresis is beyond the scope of this article. For additional details, we refer to the literature.[75-77]

G. Brownian Dynamics Simulation of Polymer Melts

Kremer, Grest, and co-workers[24,78] have undertaken a series of Brownian dynamics simulations of an off-lattice model of a polymer melt. We remind the reader that in Brownian dynamics one solves Newton's equation of motion for the system coupled to a heat bath. In their particular realization, the inertial term is retained, while in many simulations since polymers are typically in the high friction limit, it is dropped.[79] They have studied the dynamics of up to $n = 400$ chains confined to a box containing $N = 10$ polymers. Their results can be summarized as follows. They find for $n > 35$, a crossover to $D \sim n^{-2}$, scaling. For $n < 35$, $Dn \sim$ constant is seen, and $g_5(t)$ is Rouselike. As n increases further just as in the cubic lattice system, there is a crossover to $t^{1/4}$ behavior, which these authors interpret as due to reptation. Further analysis of the primitive path dynamics indicates that for the range of times sampled, the reptation component of the motion is important. These authors again interpret this as proving the existence of reptation. Unfortunately, they only sample the dynamics, at best, to the end of the $t^{1/4}$ regime, and as we have discussed in Section II.A.2, it is not until the end of this regime that the lateral component of the motion begins to dominate to the point that reptation can be neglected. Moreover, neither the existence of a $t^{1/4}$ regime nor the scaling of D, as the inverse square power of n, is unique to reptation. Thus the Kremer et al.[24,78] simulations agree with the MCD simulations of Kolinski et al.[23,47,48] in the time regime that they overlap, and the Kremer et al. case for reptation is not at all definitive.

III. THEORETICAL TREATMENTS OF POLYMER DYNAMICS

There have been a number of theoretical models developed over the years to describe the dynamics of entangled polymers. Briefly, these can be divided into three general categories. First of all, there is the classical reptation theory in which there is always a well-defined tube constraining the chain of interest.[1,4-9,13,37] This tube exists for times on the order of the terminal relaxation time of the end-to-end vector. Hence, the dominant long-wavelength motion involves the slithering down of the contour for times on the order of the terminal relaxation time. This model has a very large number of variants and is by far the most highly developed. As these models have been extensively discussed in detail elsewhere,[1] we refer the reader to the literature for a detailed exposition of their properties.

The second class of models envisions the polymer environment to be more liquid-like. Entanglements between chains are still important, but they are dynamic in nature. The models of Fujita and Einaga,[9,10] Skolnick et al.,[16,17] Ngai et al.,[14,74] and Fixman,[18,19] fall into this category. The qualitative description of each of these models is presented in turn below. Finally, there is the hydrodynamic interaction model of Phillies,[11,12] in which entanglements in concentrated polymer solutions are relegated to a minor role and interchain hydrodynamic interactions are assumed to dominate. While the theory has been developed to treat the case of concentrated solutions, since it is clearly related to the problem of dynamics in a melt, an overview of this alternative view is clearly appropriate. As will become apparent, there are a number of quite different viewpoints of the nature of polymer melt dynamics, and this, no doubt, reflects the intrinsic difficulty of the problem.

A. Fujita–Einaga Theory—The Noodle Effect

For the case of concentrated polymer solutions and polymer melts, Fujita and Einaga[9,10] argue that dense polymer systems do not suffer from the severe topological constraints conjectured by reptation theory; rather, in entangled systems, moving chains induce the movement of surrounding chains through the interactions at entanglement points, thereby producing considerable energy dissipation. These authors refer to the cooperative motion of the surrounding chains as the "noodle effect."

1. Diffusion Constant

These authors[9] then proceed to calculate the diffusion coefficient of a chain in the case of a monodisperse system in the $n/n_e \gg 1$ limit, where n_e is the mean number of monomers between adjacent entanglement points. Assuming that the translational friction coefficient of a chain can be expressed as the product of n times a mean monomeric friction coefficient ζ by the standard Stokes–Einstein relationship, the polymeric diffusion coefficient is

$$D = k_B T/(n\zeta), \tag{18}$$

with k_B Boltzmann's constant and T the absolute temperature. If the entangled segments are assumed to be localized at every n_e bead, each of which has an additional friction constant ζ_e in addition to the friction constant in the absence of entanglements ζ_0, then in the large n/n_e limit the average monomeric friction constant is

$$\zeta = \zeta_0 + \zeta_e/n_e. \tag{19}$$

Thus the crux of the model involves the calculation of ζ_e.

Fujita and Einaga make two assumptions from which ζ_e follows. First, they assume that the velocity field E, induced around the test chain when the test chain moves with velocity u, is spherically symmetric about the center of mass G, and its radius is of order $\langle S^2 \rangle^{1/2}$, the root mean square radius of gyration. In particular, they take the radius of E to equal $2\langle S^2 \rangle^{1/2}$, which is proportional to $n^{1/2}$. Second, they invoke the free drainage approximation (i.e., they neglect hydrodynamic interactions between segments) and proceed to calculate the force F in the direction u that must be applied to G to overcome the frictional force exerted by all the chains on the test chain

$$F = \zeta_0 \int_E \rho_e \bar{u} \, dv. \tag{20}$$

Here, ρ_e is the average segment density in the volume element dv which moves cooperatively with the test chain, and $\bar{u}$ is the average velocity of the coupled segments in the u direction. It is quite reasonable that ρ_e and $\bar{u}$ monotonically decrease as one goes out from the center of mass.

They replace $\bar{u}$ by ku with k a constant between zero and one that is taken to be independent of n, polymer concentration, and position within the velocity field E. In this approximation, Eq. (20) becomes

$$F = \zeta_0 ku \int_E \rho_e \, dv. \tag{21}$$

They then approximate by $N_c s$, where N_c is the number of chains directly entangling with the test chain and s is the number of segments of such an entangling chain that has substantial coupling with the test chain and which are located inside the field E.

Now the spherical domain occupied by each of the entangling chains has $(n/n_e)^{1.5} n_e$ entangling segments [there are $(n/n_e)^{1.5}$ entangling units in the volume $2\langle S^2 \rangle^{1/2}$, each contains n_e segments]. The essential assumption is then made that s is proportional to $(n/n_e)^{1.5} n_e$, with a proportionality constant that is independent of the polymer concentration. This assumption assumes that a fraction of the chains not involved in direct entanglements with the test chain are also dragged along with it.

Employing these assumptions, Eqs. (18–21) give

$$\zeta_e = \zeta_0 f n^{3/2} n_e^{-1/2} \tag{22}$$

with f a constant. Thus the effective monomeric friction coefficient is

$$\zeta = \zeta_0 [1 + f(n/n_e)^{3/2}]. \tag{23a}$$

When ζ is inserted in Eq. (18), the self-diffusion coefficient is

$$D^{-1} = n\zeta_0(k_B T)^{-1}[1 + f(n/n_e)^{3/2}]. \tag{23b}$$

Equation (23b), therefore predicts that D asymptotically should be proportional to $n^{-5/2}$, a behavior not inconsistent with experiment.[9,80]

One of the touchstones of a successful theory is the ability to rationalize not only the molecular weight dependence of a monodisperse polymer melt, but also to rationalize the independence of the probe diffusion coefficient on matrix molecular weight when the latter is sufficiently large. In their treatment Fujita and Einaga[9] consider a bidisperse blend of polymers having degree of polymerization n_1 and n_2 for components 1 and 2. Both components are assumed to be sufficiently long that they exhibit entangled behavior and $n_2 < n_1$. The key to this analysis lies in the assumptions indicated above that only chains within several (two) radii of gyration of the chain of interest are dragged along with the probe chain. This leads to the prediction that the increase in friction coefficient is identical to Eq. (22), with n_2 substituted for n, and where n_e is given by the reciprocal average of the mean number of monomers between entanglements arising from chains 1 and 2. Thus the desired independence of matrix molecular weight is recovered with $D_p \sim n_2^{-5/2}$. However, fits to experiment did not reveal the theoretically predicted power of D_p.

Einaga and Fujita[9] have also been able to derive a prediction of the concentration dependence of D, which obtains from the recognition that n_e should be inversely proportional to polymer concentration c_p. This simply gives $D \sim c_p^{-1.5}$. Experimental data give varying power law (or perhaps not; see the Phillies theory,[11,12,34] Section III.D) dependencies of D, ranging from the -3 to -1.75 power. However, different c_p scaling can be obtained depending on the molecular-weight dependence assumed.

2. Viscosity

Fujita and Einaga[10] have also developed an expression for the steady-state shear viscosity of a polymer blend, η_0. They basically employ standard linear viscoelasticity theory plus the assumption that the terminal relaxation time of each component τ_{mi} is given by

$$\tau_{mi} = Bn_i/D_i, \tag{24}$$

where B is a constant and D_i is the diffusion coefficient of the ith component of degree of polymerization n_i. In the case of a monodisperse system, it then follows that η_0 is predicted to be proportional to the 3.5 power of the molecular weight. By assuming that the shorter chains relax as if they were in a noninteracting blend, they also derive an expression for η_0 appropriate to this situation.

The key point is that this theory predicts for a monodisperse system that $D\eta_0 \sim n$. Furthermore, this analysis shows how an alternative mechanism to reptation based on the assumption that what is relevant is the dragging of chains for distances on the order of the radius of gyration, can give rise to a plausible physical picture that overall is in fairly good agreement with experiment. It is interesting to note that reasonable scaling behavior can be predicted without, in fact, specifying whether the chains ultimately disengage by reptation, or perhaps by some other mechanism.

B.   Phemonenological Theory of Dynamic Entanglements

1.   Diffusion Constant

Based on the discussion above, the question arises as to just how general the treatment must be to recover the experimentally observed scaling of D and η with n. The following development by Skolnick et al.[16,17] is patterned after the Hess[13,81] generalized Rouse treatment, which was used to provide a theoretical underpinning for reptation-like behavior. The key to Hess' recovery of reptation is the assertion that the forces exerted on the test chain by the sea of surrounding chains act perpendicular to the chain axis. Since the Kolinski et al.[23,47,48] simulations indicated that this is only true for a small fraction of time in the intermediate time regime, we modified the treatment to account for the observed isotropic motion of the chains. In particular, because the behavior of the chains is Rouse-like, the motion is factored into two components, the center of mass motion between the chains and the motion of the internal Rouse-like coordinates, with a weak coupling between the two permitted.

Starting from the Green–Kubo expression for the diffusion constant, Hess employed Zwanzig–Mori projection operator techniques to obtain the enhanced friction constant acting on the test chain due to the other chains.[13] After a bit of arithmetic,

$$D = \frac{k_B T}{n[\zeta_0 + \int_0^\infty dt\Delta\zeta(t)]}. \tag{25}$$

Here ζ_0 is a generalized concentration-dependent Rouse monomer friction coefficient, and $\Delta\zeta$ is the additional dynamic friction term arising from interchain interactions. If a standard separation of time scales argument is made, the effects of short-lived contacts can be subsumed into an effective molecular weight independent, monomeric friction coefficient ζ_0. The effect of the long-lived, but dilute dynamic entanglements between chains are reflected in $\Delta\zeta$, which is explicitly defined as

$$\Delta\zeta(t) = \frac{\langle F(t) \cdot F(0) \rangle}{3nk_B T} \tag{26}$$

where the term in the brackets is a Zwanzig–Mori projected force correlation function.

In order to evaluate $\Delta\zeta$, the following assumptions are made. First, interactions between chains are assumed to be predominantly steric in nature and hydrodynamic interactions are ignored. Second, the time evolution of $\Delta\zeta$ reflects the motion of the dynamic entanglements. Finally, because dynamic entanglements are dilute, the interaction hierarchy can be truncated at the pair level. Using these assumptions, it is straightforward to show that the propagator for pairs of chains in contact in the long-wavelength, hydrodynamic limit is of the form

$$R(q, t) = \exp(-D_{\text{eff}} q^2 t) \tag{27}$$

where D_{eff} is an effective diffusion constant and q is the magnitude of the wave vector.

A plausible functional form for D_{eff} is given by

$$D_{\text{eff}} = (1 - \beta)D_0 + \beta D. \tag{28}$$

D_0 is the renormalized Rouse diffusion coefficient given by

$$D_0 = k_B T / n\zeta_0 = d_0 / n. \tag{29}$$

Equation (28) accommodates the fact that for times of order of the terminal relaxation time the behavior is essentially Rouse-like (see Sections II.A and II.D.1), but with a small coupling between the center of mass motion and the internal coordinates. In the small coupling limit ($\beta \ll 1$), it follows that

$$D = \frac{d_0}{n(1 + n/n_e)}. \tag{30}$$

Here n_e is the average number of monomers between dynamic entanglements.

An analogous equation can be derived for the diffusion coefficient of a probe in a matrix, which in the small coupling limit gives

$$D_p = \frac{d_0}{n_p}\left[\frac{1 + n_p/n_m}{1 + n_p/n_m + 2n_p/n_e}\right] \tag{31}$$

where n_p (n_m) is the degree of polymerization of the probe (matrix) chains.

Equation (30) and (31) provide rather good fits to simulation results described in Sections II.A–II.C as well as to the experimental data of Antonietti et al.[67] on polystyrene melts and predict that D_p should scale asymptotically as n_p^{-2}. A more detailed description of the model that describes the behavior of the self-diffusion constants over the entire range of β is given elsewhere.[16,49]

Since the assumptions employed to derive Eqs. (30) and (31) are so benign, the existence of an n^{-2} power law dependence of D cannot be used to prove the existence of any particular microscopic model of chain motion. It differs from the Fujita–Einaga result[9] in that the latter assumes stronger coupling between entangled chains, and here a weak coupling limit is assumed.

2. Viscosity

We next review those features that any successful theory must rationalize about the internal dynamics of polymer melts. It must reproduce the molecular weight dependence of the diffusion coefficient and the viscosity.[1,26–33,35,67] It must also be consistent with the simulation results[23,47,48] that indicate that the chain motion is slowed down and in many ways Rouse-like and that there is no tube confining the chains. The single bead autocorrelation function has a t^b regime with $b < 1/2$, and g_{cm} has a t^a regime with $a < 1$. The product $D\tau_R/n$ scales as n^ε with ε, assuming values between 0.1 and 0.2. Finally, it must rationalize the single-bead mean square displacement profiles which indicate that the ends are more mobile than the equivalent Rouse chain and the middle is less mobile (see Fig. 12).

We summarize below the features of a recent phenomenological theory that accounts for the above features. The following simplifying assumptions are made. (i) At short times, a la Doi and Edwards,[5] the response of the melt is treated as rubber. We then focus on the motion of an average reporter chain and assume that the long-time relaxation behavior in a polymer melt is adequately described by a Rouse model; however, because of the presence of dynamic entanglements there are n/n_e slow moving points, each of which drags another chain along with it. In the effective single-particle picture, these slow moving points have an augmented monomer friction constant of order n. This approximation assumes that for times on the order of the terminal relaxation time, on average the matrix chains are dragged along with the test chain and the dissolution of the contact can be ignored. This clearly is the simplest approximation of the effect of entanglements. It neglects the coupling between the various entanglements and the time course of their dissolution and formation.

With these assumptions in hand, the following qualitative picture of the dynamics emerges for the expected crossover behavior of the individual chains. Suppose an individual chain has just one entanglement. Physically, one would expect this entanglement to be located at or near the center of the chain.

Clearly, this one slow-moving point wouldn't change the behavior of the terminal relaxation time of the end-to-end vector by much; the single slow-moving point behaves like a local defect. However, in the absence of the entanglement, the self-diffusion coefficient is d_0/n, and in the presence of the entanglement, it is $d_0/2n$. In other words, the crossover behavior of D and τ_R should be different, in agreement with experiment. Moreover, the center of

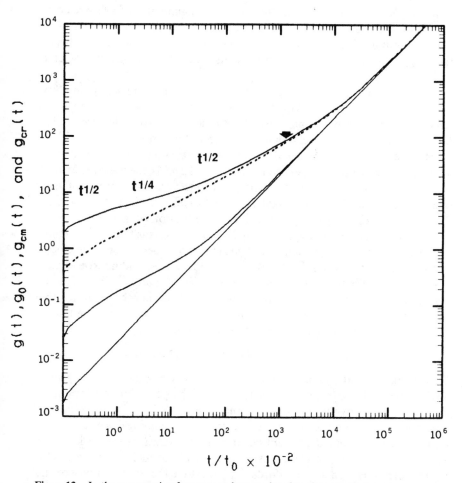

Figure 13. In the curves going from top to bottom, log–log plots of $g(t)$, the mean square displacement per bead obtained assuming that each bead has a uniform friction constant $g_0(t)$, the mean square displacement of the center of mass $g_{cm}(t)$, and the mean square displacement of the center of frictional resistance $g_{cr}(t)$; t_0 is the average time a monomer takes to diffuse a bond length. See text for further details.

mass and the center of frictional resistance are no longer the same.[42] Basically, the center of mass motion couples into the internal coordinates, and this gives rise to the $g_{cm}(t) \sim t^a$, for values less than $2\langle S^2 \rangle$. Similarly, a t^b regime is predicted for $g(t)$.

It is also possible to show that in the limit as n goes to infinity, the dynamic properties behave as if each monomer has an effective friction constant $\zeta_0(1 + n/n_e)$ with ζ_0, the monomeric friction coefficient in the absence of chain connectivity. The viscosity equals 4/15 of the Doi–Edwards[37] value, and ultimately scales as n^3. Thus, in the crossover regime, the viscosity scales as $n^{1+\delta}$ with δ going to zero, as n goes to infinity. Finally, the product of the plateau modulus times the shear compliance equals 10/7, while Doi–Edwards theory gives a value of 6/5, and experiments give values in the range of 2.5 to 3.[37]

Fits to a number of assumed distributions of slow-moving points give $\eta \sim n^{3.4}$, with a lower bound for the crossover value of $\eta \sim n^3$, around 40–50 entanglements. This is not inconsistent with the recent work of Colby et al.,[35] but the experiments themselves are the subject of controversy.

The present theory predicts that the crossover value of n'_c, is about $4.5n_c$. Thus, in accord with the experiment[35] (see Eqs. 1 and 2), the viscosity exhibits the 3.4 power of the molecular weight, prior to the crossover of the diffusion constant into the n^{-2} regime. Finally $D \tau_R/n$ has an n^ε regime with $\varepsilon = 0.1$ or 0.2, depending on the particular distribution of friction constants.

Finally, in Fig. 13 log–log plots of $g_{cm}(t)$ and $g(t)$ versus time are presented for the case of an $n = 255$ chain with a mean distance between entanglements of 15 and a τ_R of 1.88×10^5. This τ_R value of the corresponding Rouse chain is 1.3×10^4. In the top solid curve, there are $t^{1/2}$, $t^{1/4}$, and $t^{1/2}$ regimes in $g(t)$, and these chains are clearly not reptating. This is but another example that shows that the existence of a $t^{1/4}$ regime in $g(t)$ is indicative of some kind of constrained dynamics, but it need not be reptation.

C. Coupling Model of Polymer Dynamics

Over the past decade Ngai, Rendell, and co-workers, have developed a general formalism to address the problem of how[14, 74, 82] the relaxation of a "primitive" mode is modified by the coupling to complex surroundings. The fundamental prediction of the model is that the net effects of such coupling can be accounted for by the use of a time-dependent relaxation rate. As will be seen below, their application to polymer dynamics is in the same philosophical spirit as that of the two previous models, but affords the advantage that it explicitly accounts for the time-dependent modification of the behavior of the chain.

The coupling model asserts that the coupling of the primitive relaxation mechanism to the environment slows down the primitive relaxation rate W_0 for times greater than a characteristic time $t_c = \omega_c^{-1}$. In other words, there is a time-dependent relaxation rate $W(t)$ of the form

$$W(t) = W_0 \qquad \omega_c t < 1, \tag{32}$$

$$W(t) = W_0(\omega_c t)^{-n} \qquad \omega_c t > 1.$$

The relaxation function $c(t)$ then obeys the rate equation

$$\frac{dc}{dt} = -W(t)c, \tag{33}$$

and the normalized relaxation function $\phi(t) = c(t)/c(0)$ is of the form

$$\Phi(t) = \exp(-t/\tau^*)^{1-n}, \tag{34}$$

and the terminal relaxation time is

$$\tau^* = [(1 - n)\omega_c^n \tau_0]^{1/(1-n)}. \tag{35}$$

The degree of coupling is embodied in the parameter n (not to be confused with the degree of polymerization; we employ the Ngai et al.[14,74,82] notation to facilitate comparison with the original papers). For chain molecules in the melt, τ_0 is the terminal relaxation time of the nonentangled Rouse chain. The particular value of n is taken to depend on the particular kind of experimental measurement. For the case of shear viscoelasticity, they report the following predictions. The terminal dispersion of the shear modulus is of the Kohlraush, et al. form,[83] and the terminal relaxation time can be calculated from τ_0 using Eq. (35) for $\Phi(t)$. The former prediction is in accord with a number of experiments.[14] Thus, using Eq. (35) and the Rouse τ_0, τ^* can be extracted. These fits provide a values of n_η in the range of 0.4 to 0.45. If one then approximates the viscosity by the relation $G_N^0 \tau^*$ with G_N^0, the plateau modulus, then

$$\tau^* \sim M^{2/(1-n_\eta)} [\zeta_0(T)]^{1/(1-n_\eta)}. \tag{36}$$

In Eq. (36), M is the molecular weight of the polymer. Using the values of n_η obtained from the experiment gives a molecular weight dependence of the viscosity between 3.3, and 3.6, in good agreement with experiment. Equation (36) also provides an estimate of the temperature dependence of τ_η^* that is in agreement with experimental measurements on polyethylene and hydrogenated polybutadiene. Finally Eq. (36) provides a relationship between the viscosity activation energy E_η^* and the monomeric friction constant activation energy E_a,

$$E_\eta^* = E_a/(1 - n_\eta), \tag{37}$$

a relationship also found to agree with the experiment.[14]

They[84] have also deduced the molecular weight dependence of the diffusion constant. They proceed by relating the observed activation energy of the diffusion constant E_D^* to the monomeric diffusion constant by the analogous relation, as in Eq. (37). This provides a value of n_D of 0.33, and a characteristic relaxation time associated with diffusion of $\tau_D^* = M^{2/(1-n_D)}$ which scales as M^3. By assuming that $D \sim \langle S^2 \rangle / \tau^*$, they recover the inverse squared power dependence of D on M.

More recently, Ngai et al.[85,86] have also derived a relationship for the molecular weight dependence of a probe chain dissolved in a matrix that is in good agreement with experiment for both linear and star tracer molecules. Furthermore, they have obtained the stretched exponential form for the concentration dependence of D (see Eq. 2). They identify the origin of the stretched exponential with the gradual increase of the coupling between a diffusant and the matrix as the concentration of the diffusant increases; that is, the coupling parameter $n(c)$, monotonically increases with increasing concentration c.

In a recent publication Ngai and Lodge[86] applied the coupling model to treat the diffusion of 3- and 12-arm polystyrene stars in entangled PVME solutions. They find excellent agreement with experiment. Moreover, their treatment predicts a different temperature dependence for stars than the corresponding linear chains, a prediction that is also in accord with experiment. Thus, in application to extant experimental data, the coupling model is in rather good agreement with experiment, without invoking the existence of reptation. The qualitative picture is also in accord with simulation data[18,23,47,48,69-71] that does not provide support for the existence of reptation within a spatially fixed tube. The only problem with the theory is that it cannot, as yet, predict the values of the coupling parameters, nor does it identify the particular mechanism responsible for the coupling. Nevertheless, in spite of these limitations, the coupling model is a very powerful approach that appears to have much to say about the dynamics of entangled polymer systems.

D. The Fixman Model of Polymer Melt Dynamics

In a recent pair of papers, Fixman[18,19] has proposed a generalization of the reptation model of polymer dynamics that has as its basis the following assumptions. First, it assumes that the most naive view of both simulations and experiments are correct. That is, the viscosity $\eta \sim n^{3.4 \pm 0.1}$ is indeed the asymptotic power law, that the highly cooperative and simultaneous motion of the chains, seen in simulations,[17,23,47,48,61] is correct at all chain lengths (i.e., there is no frozen matrix) and, finally, that the translational friction coefficient for self diffusion scales as n^2. As will be seen below, unlike the treatments discussed above that in many respects are highly phemoneno-logical, Fixman has attempted to identify the specific mechanisms responsible for entangled melt behavior and proposes a formalism for calculating the polymeric properties arising from this conjectured behavior.

Taking these assumptions as valid implies that chains can diffuse many radii of gyration without orientational relaxation.[35] The persistence of entanglements during the process of diffusion across many radii of gyration is rationalized in terms of the correlated reptative motion of the probe and vicinal (matrix) chains. The mathematical formalism is cast in terms of generalized Langevin equations for the motion of the probe and has as its basis the Rouse–Zimm model of chain dynamics at infinite dilution. In particular, Fixman assumes that the matrix surrounding the probe chain is capable of exerting a viscoelastic response to any forces exerted on it. But unlike the isolated chain case where the elastic part is essentially neglected, in the case of melts, the elastic part is assumed to be important. In addition to the viscoelastic response, the matrix chains are themselves capable of reptative motion along their respective primitive paths. Fixman asserts that for the consistency of the treatment of viscoelastic interactions, the reptative motion of the probe chain is accompanied by reptative motion of the matrix chains; that is, unlike in standard reptation theory, here the motion of the probe and matrix chains is taken to be highly correlated. Entanglements are identified as small loops or twists of one chain about the other.

Imagine then the response of the polymeric system to an external force. Owing to the elastic deformation of the matrix, the displacement of the probe chain with respect to the laboratory fixed frame increases, concomitantly, the reptative diffusion relative to the deforming matrix slows down. The net result on the translational diffusion of the probe of these two opposing effects is predicted to be negligible, if both the probe and vicinal chains are of the same length. However, unlike standard reptation theory, the friction constant for reptative motion increases by a factor $n^{1-x_s} \cdot x_s$ assumes the value of one-half, if Gaussian statistics obtains and 0.6 if excluded volume statistics applies. Furthermore, the translational diffusion constant $\sim n^{-2}$, and the viscosity scales like n^{4-x_s}.

In a companion paper[19] Fixman applied his formalism to calculate the storage and loss modulus, the shear compliance, and the translational diffusion coefficient using parameters appropriate to polybutadiene. The resulting dynamic stress modulii are in fairly good agreement with experimental results. A particularly important conclusion is that the Langevin equation has the same normal mode structure as in Rouse–Zimm[42] theory, but each mode relaxes as a sum of exponentials, rather than just a single exponential (perhaps providing the underpinnings for the Ngai et al.[14, 74, 82, 84–86] coupled model, KWW functional form). Fixman points out that there his approach requires a number of not-well-determined parameters; nevertheless, the theory is rather impressive in its agreement with experiment.

At this junction it is appropriate to summarize the various viewpoints. In all the theories discussed above, there is the implicit assumption of the cooperative relaxation of the test chain with its surrounding matrix of chains. That is, the motion of the probe and matrix chains are coupled, and there is no fixed matrix confining the chain of interest. Furthermore, the entanglements are envisioned to be of a dynamic nature, and not static as in the classical reptation model. In fact, the various theories differ in their specification of what mechanism, if any, is responsible for entanglement disengagement. In all but the Fixman theory, the disengagement mechanism is left unspecified. Fixman asserts that the dominant long-time disengagement mechanism is by reptation. Furthermore, all the other theories implicitly or explicitly assume that the decay distance for orientational correlation is on the order of the radius of gyration of a chain, whereas Fixman envisions this as occurring over many radii of gyration. What is clear, however, from the discussion of all the aforementioned theories is that (more or less) they all reproduce most, if not all, features of experiment, and more sophisticated tests will have to be devised to determine which, if any, are in fact correct.

E.   Hydrodynamic Interaction Theory of Concentrated Solutions— The Phillies Model

For the case of concentrated polymer solutions, Phillies[34] has found that the stretched exponential form for the self-diffusion coefficient,

$$D = D_0 \exp(-\alpha c^v), \tag{38}$$

fits the entire concentration range for all extant studies of the concentration dependence of D. In fact, the stretched exponential form fits the concentration dependence of both high- and low-molecular-weight polymers, as well as for globular, nonentangling proteins. Since globular proteins cannot entangle (they are rigid, dense bodies), Phillies[34] argues that the same physics underlying the motion of dynamics of globular proteins should hold for polymer solutions as well. Unlike the Ngai and Lodge treatment described above[86] (see Section III.C). Phillies has developed a model that gives the functional form of Eq. (38), and numerical values of α and v.[11, 12] Thus, more than one physical picture can reproduce Eq. (38), and in fact Adler and Freed[87] have derived Eq. (38) in the context of a mean field approximation. The Phillies approach is especially useful in that it provides for quantitative values for all of the parameters.

The fundamental assumptions of the Phillies model[11] are: (1) There is a self-similar effect of infinitesimal concentration increments on D. (2) The dominant force between polymers in semidilute solution is hydrodynamic. (3) The root mean square radius of gyration is described by the blob model of Daoud et al.[88] that is, $\langle S^2 \rangle \sim Mc^x$ with $x = -1/4$. An overview of his derivation of Eq. (38) follows.

From assumption (1), the increase in the mobility μ arising from a concentration increment Δc is taken to be

$$\mu(c + \Delta c) = \mu(c) + A\Delta c, \qquad (39)$$

where A is assumed to be proportional to $\mu(\Delta c)$. The dependence of A on polymer dimensions is taken from hydrodynamics by employing the functional form appropriate to interactions between hard spherical particles, namely, $A \sim \langle S^2 \rangle^{1/2}_a \langle S^2 \rangle^{3/2}_b$, and where a refers to the test polymer and b refers to the other chains in solution.

The key to the argument comes next, where assumption (3) is invoked; namely,

$$\frac{d\mu}{\mu} = a(c)\, dc, \qquad (40)$$

with $a(c) = A/\mu$. By examining the functional form for μ, as a function of c, Phillies replaces A/μ by A/μ_0, with μ_0 the mobility at infinite dilution. Integrating Eq. (40) and using the Stokes–Einstein relation between D and μ gives

$$D(c) = D_0 \exp\left(\int_0^c \frac{A(c)}{\mu_0}\, dc \right). \qquad (41)$$

If the scaling of $\langle S^2 \rangle$ of the probe and matrix with concentration is used, one finds that

$$\int dc\, \frac{A(c)}{\mu_0} = \alpha c^v \qquad (42)$$

where $\alpha \sim M$ and $v = 1 - 2x$. An analogous development of Eqs. (40) and (41) has been previously presented by Adler and Freed.[87]

In subsequent work, in the spirit of the hydrodynamic interaction model of concentrated polymer solutions, Phillies[12] derived an explicit functional form for $A(c)$. The specific approach is based on a generalization of Einstein's derivation of the Stokes–Einstein diffusion equation. In particular, the method of reflections is used where a series of velocities are calculated. The first polymer is assumed to have an unperturbed mobility. This then acts on the second polymer, which creates a velocity echo that perturbs the original polymer's flow field and consequently its mobility. The treatment, in detail, involves the generalization of the Kirkwood Riseman[89] model from one to two polymers. The net result is that

$$\alpha = -\frac{9}{16}\frac{\langle S^2 \rangle_a^{1/2}}{2a_0}\frac{4\pi}{3}\frac{\langle S^2 \rangle_b^{3/2}}{M_b}\frac{1}{1 - 2x},$$
(43)

where a_0 is the radius of a monomer. Note that Eq. (43) predicts that $\alpha \sim (M_a M_b)^{1/2}$ in agreement with previous scaling arguments.[11]

For the case of polystyrene at $M = 1 \times 10^6$, Phillies estimates α is about -2. Fits to experiment give an α of -0.7. However, over the entire range of α, Phillies reports rather good agreement with the experiment, with α ranging over more than 3 orders of magnitude as M changes.

Phillies further argues that α should be similar for both linear and star molecules, which appears to be in agreement with experiment.[12] In a recent pair of papers,[90,91] however, Lodge et al. find that D_0 values fit to linear polystyrene data are systematically greater than the measured values and that the expected scaling of α with molecular weight is not found, in contrast to previous results.[34] For the case of stars, Eq. (38) reproduces the concentration dependence very well, but the origin of the chain architecture dependence is unclear. Thus, a controversy between the Lodge and Phillies groups is unresolved.

IV. SUMMARY AND CONCLUSIONS

Since the early 1970s when it was first proposed, the reptation model has been the most popular and widely accepted description of dynamics in dense polymer systems.[1,4-8,37] Undoubtedly, it is a very elegant and simple model that can rationalize a number, but by no means all, of the experimental observations. Because of its relative maturity, it has been applied to a number of experimental situations. It is appropriate here to review its successes and failures. The model can reproduce the molecular weight dependence of the self-diffusion coefficient of linear chains, and it provides a scaling of the shear viscosity that goes like the cube of the molecular weight, whereas experiment indicates that it goes like the 3.4 power of the molecular weight. It predicts a $t^{1/4}$ regime in the mean square displacement profile of single beads, a regime seen in two simulations on linear chains. It predicts that the diffusion coefficient of a probe in a linear matrix should be independent of matrix molecular weight, that rings should move substantially slower than linear chains, and that stars should move exponentially slower in the number of arms than the corresponding linear chains. The reduction in mobility of stars predicted by the reptation theory has been taken to vindicate the reptation/arm retraction mechanism.[90,91] However, the viscosity of rings[39,40] is in fact smaller than the corresponding linear chains, and granted that while the rings do not span as broad a molecular weight range as linear chains, they are in the 3.4 power of M regime for η, and it is difficult to understand why they are not moving slower if reptation, in fact, dominates.

In the present review we have summarized results from alternative theories that do not invoke the existence of a fixed tube, and yet the molecular weight dependence of D is recovered. In fact, the requirements for the $D \sim n^{-2}$ and the asymptotic matrix molecular weight independence of a probe diffusion constant follow from such weak considerations that no specific microscopic mechanism of D can be inferred from its molecular weight dependence.[16] Similarly, the existence of $t^{1/4}$ regimes in systems that cannot possibly reptate[17,65] shows once again that this too is a signature of cooperative dynamics and nothing more. With respect to different chain topologies, the analytic models are less developed, but just because a model isn't mature, it shouldn't be dismissed out of hand. A number of simulations on rings and linear chains do not find clear evidence for the existence of a fixed tube.[17,23,47,48,59,68,70,71] While there is some memory of the initial configuration (even free Rouse chains exhibit this at short enough times), as time increases for distances and times where the tube should be well-defined if reptation is correct (into the second $t^{1/2}$ regime of the single-bead autocorrelation function), the reptation component becomes of minor importance, and the environment surrounding the chains is not static; rather, cooperative back-flow effects are evident. This is true for both dynamic Monte Carlo[23,47,48,59,68,70,71] and Brownian dynamics simulations,[18] although this viewpoint has been questioned.[24,78]

These models, which treat the surroundings of a chain as a more fluid-like, and which view entanglements as being of a dynamic nature (although the exact mechanism is not understood), are equally successful as the reptation model in predicting the experimental phenomenology of linear chains. Moreover, these models agree with the simulation results. If one views the ability to treat the effects of different topology as the test of the validity of the theory, then reptation fails the test for ring melts. The development of alternative theories for ring and branched molecules is required before these theories can be dismissed or fully accepted. While the coupling model of Ngai[86] has been applied to stars, since it makes no attempt at specifying the precise microscopic mechanism of entanglements, it has been argued that it is not capable of verifying the dynamic entanglement picture. Undoubtedly this is true, but once again, if a general class of models fits the data, this implies that until such time as features peculiar to a particular motional mechanism are identified, then it is impossible to confirm the validity of a particular model.

In summary, while the problem of dynamics in dense polymer systems has been the subject of study for the past 40 years, the precise mechanism of motion is not in fact resolved. A number of theories having quite a different physical basis can reproduce the experimental results, and because the simulation of a system well into the entangled regime is likely to remain beyond conputational capabilities for considerable time to come, the final disposition of the competing theories will not be fully resolved by recourse to simulation alone. (It can always be argued that if the desired reptation behavior is not seen, then the chains are too small.) As alternative analytic theories to reptation are more fully developed and applied to different chain topologies, and their predictions tested against experiment and simulation, then the controversy about whether reptation in a fixed tube is valid or not will hopefully be resolved. This review has attempted to make the case that the matter is not at all settled in favor of reptation in a fixed tube, and that an abundance of evidence exists that argues against the simple reptation model's validity. However, whether or not reptation ultimately proves valid, it was the first model that successfully rationalized a wide body of experimental data and provided a conceptual basis for the design of new experiments. In this alone, it has proven to be extremely valuable, and its importance to the field of polymer dynamics cannot be overemphasized.

Acknowledgment

The authors gratefully acknowledge the support of a grant from the polymer program of the National Science Foundation, No. DMR 85-20789. Useful discussions with Dr. K. Ngai, Dr. George Phillies and Dr. A. Sikorski are acknowledged.

References

1. W. W. Graessley, *Adv. Poly. Sci.* **46**, 67 (1982).
2. F. Bueche, *Physical Properties of Polymers*, Wiley, New York, 1962.
3. G. C. Berry and T. B. Fox, *Adv. Poly. Sci.* **5**, 261 (1968).
4. P. G. de Gennes, *J. Chem. Phys.* **55**, 572 (1971).
5. M. Doi and S. F. Edwards, *J. Chem. Soc.* **74**, 1789 (1978).
6. M. Doi and S. F. Edwards, *J. Chem. Soc.* **74**, 1802 (1978).
7. M. Doi and S. F. Edwards, *J. Chem. Soc.* **74**, 1818 (1978).
8. M. Doi and S. F. Edwards, *J. Chem. Soc.* **75**, 38 (1978).
9. H. Fujita and Y. Einaga, *Poly. J.* (*Tokyo*) **17**, 1131 (1985).
10. H. Fujita and Y. Einaga, *Poly. J.* (*Tokyo*) **17**, 1189 (1985).
11. G. Phillies, *Macromolecules* **20**, 558 (1987).
12. G. Phillies, *Macromolecules* **21**, 3101 (1987).
13. W. Hess, *Macromolecules* **19**, 1395 (1986).
14. R. W. Rendell, K. L. Ngai, and G. B. McKenna, *Macromolecules* **20**, 2250 (1987).
15. M. Fixman, *J. Chem. Phys.* **89**, 3892 (1988).
16. J. Skolnick, R. Yaris, and A. Kolinski, *J. Chem. Phys.* **86**, 1407 (1988).
17. J. Skolnick and R. Yaris, *J. Chem. Phys.* **86**, 1418 (1988).
18. M. Fixman, *J. Chem. Phys.* **89**, 3892 (1988).
19. M. Fixman, *J. Chem. Phys.* **89**, 3912 (1988).

20. A. Baumgartner, *Ann. Rev. Phys. Chem.* **35**, 419 (1984).
21. M. Bishop, D. Ceperley, H. L. Frisch, and M. M. Kalos, *J. Chem. Phys.* **76**, 1557 (1982).
22. M. T. Gurler, C. C. Crabb, D. M. Dahlin and J. Kovac, *Macromolecules* **36**, 398 (1983).
23. A. Kolinski, J. Skolnick, and R. Yaris, *J. Chem. Phys.* **86**, 1567 (1987).
24. K. Kremer, G. S. Grest, and I. Carmesin, *Phys. Rev. Lett.* **61**, 566 (1988).
25. W. W. Graessley, *Adv. Poly. Sci.* **16**, 1 (1974).
26. J. D. Ferry, *Viscoelastic Properties of Polymers*, Wiley, New York, 1980.
27. P. F. Green, P. J. Mills, C. J. Palmstrom, J. W. Mayer, and E. J. Kramer, *Phys. Rev. Lett.* **53**, 2145 (1984).
28. P. F. Green and E. J. Kramer, *Macromolecules* **19**, 1108 (1986).
29. M. Antonietti, J. Coutandin, R. Grutter, and H. Sillescu, *Macromolecules* **17**, 798 (1984).
30. M. Antonietti, J. Coutandin, and H. Sillescu, *Macromolecules* **19**, 793 (1986).
31. M. Tirrell, *Rubber Chem. Tech.* **57**, 523 (1984).
32. C. R. Bartels, B. Crist, and W. W. Graessley, *Macromolecules* **17**, 2702 (1984).
33. H. Yu in M. Nagasawa, *Molecular Conformation and Dynamics of Macromolecules in Condensed Systems (Studies in Polymer Science, Vol. 2)*, Elsevier, Amsterdam, 1988, p. 107.
34. G. Phillies, *Macromolecules* **19**, 2367 (1986).
35. R. H. Colby, L. J. Fetters, and W. W. Graessley, *Macromolecules* **20**, 2226 (1987).
36. P. J. Flory, *Faraday Discuss. Chem Soc.* **68**, 14 (1979).
37. W. W. Graessley, *J. Poly Sci. Poly Phys. Ed.* **18**, 27 (1980).
38. J. Klein, *Macromolecules* **19**, 105 (1986).
39. G. B. McKenna, G. Hadziioannou, P. Lutz, G. Hild, C. Strazielle, C. Straupe, and A. J. Kovacs, *Macromolecules* **20**, 498 (1987).
40. J. Roovers, *Macromolecules* **21**, 1517 (1988).
41. P. Rouse, *J. Chem. Phys.* **21**, 1272 (1953).
42. H. Yamakawa, *Modern Theory of Polymer Solutions*, Harper and Row, New York, 1968.
43. K. E. Evans and S. F. Edwards, *J. Chem. Soc. Faraday Trans. 2* **77**, 1891 (1981).
44. K. E. Evans and S. F. Edwards, *J. Chem. Soc. Faraday Trans. 2* **77**, 1929 (1981).
45. S. F. Edwards and K. E. Evans, *J. Chem. Soc. Faraday Trans. 2* **77**, 1913 (1981).
46. P. G. de Gennes, *Scaling Concepts in Polymer Physics*, Cornell University Press, Ithaca, New York, 1979.
47. A. Kolinski, J. Skolnick, and R. Yaris, *J. Chem. Phys.* **86**, 7164 (1987).
48. A. Kolinski, J. Skolnick, and R. Yaris, *J. Chem. Phys.* **86**, 7174 (1987).
49. J. Skolnick, R. Yaris, and A. Kolinski, *Int. J. Mod. Phys. B.* **3**, 33 (1989).
50. A. Baumgartner and M. Muthukumar, *J. Chem. Phys.* **87**, 3082 (1987).
51. M. Muthukumar and A. Baumgartner, *Poly. Preprints* **30**, 99 (1989).
52. A. Baumgartner in *Applications of the Monte Carlo Method in Statistical Physics*, Springer-Verlag, Heidelberg, 1984.
53. K. Kremer, A. Baumgartner, and K. Binder, *J. Phys. A: Math Gen.* **15**, 2879 (1982).
54. H. J. Hilhorst and J. M. Deutch, *J. Chem. Phys.* **63**, 5153 (1975).
55. H. Boots and J. M. Deutch, *J. Chem. Phys.* **67**, 4608 (1977).
56. C. Stokely, C. C. Crabb, and J. Kovac, *Macromolecules* **19**, 860 (1986).
57. K. Binder in *Monte Carlo Methods in Statistical Physics*, Springer, Berlin, 1986.
58. P. H. Verdier and W. H. Stockmayer, *J. Chem. Phys.* **36**, 227 (1962).
59. J. Skolnick, A. Kolinski, A. Sikorski, and R. Yaris, *Poly. Preprints* **30**, 79 (1989).
60. M. Dial, K. S. Crabb, C. C. Crabb, and J. Kovac, *Macromolecules* **18**, 2215 (1985).
61. C. C. Crabb and J. Kovac, *Macromolecules* **18**, 1430 (1985).
62. K. Kremer, *Macromolecules* **16**, 1632 (1983).
63. A. Kolinski, J. Skolnick, and R. Yaris, *J. Chem. Phys.* **84**, 1922 (1986).
64. P. Romiszowski and W. H. Stockmayer, *J. Chem. Phys.* **80**, 485 (1984).

65. M. Milik, A. Kolinski, and J. Skolnick, *J. Chem. Phys.*, submitted.
66. S. F. Edwards, *Polymer* **6**, 143 (1977).
67. M. Antonietti, K. J. Folsch, and H. Sillescu, *Makromol. Chem.* **188**, 2317 (1987).
68. A. Sikorski, A. Kolinski, J. Skolnick, and R. Yaris, manuscript in preparation.
69. T. Pakula, *Macromolecules* **20**, 679 (1987).
70. T. Pakula and S. Geyler, *Macromolecules* **20**, 2909 (1987).
71. T. Pakula and S. Geyler, *Macromolecules* **21**, 1670 (1987).
72. M. E. Cates and J. M. Deutsch, *J. Phys.* **47**, 2121 (1986).
73. F. Bueche, *J. Chem. Phys.* **20**, 1959 (1952).
74. K. L. Ngai, R. W. Rendall, A. K. Rajagopal, and S. Teiter, *Ann. N.Y. Acad. Sci.* **484**, 150 (1985).
75. S. D. Levene and B. H. Zimm, *Science* **245**, 396 (1989).
76. L. S. Lehrman and H. Frisch, *Biopolymers* **21**, 995 (1982).
77. O. J. Lumpkin and B. H. Zimm, *Biopolymers* **21**, 2315 (1982).
78. K. Kremer, G. S. Grest, and B. Duenweg, *Poly. Preprints* **30**, 43 (1989).
79. E. Helfand, *Bell Syst. Tech J.* **58**, 2289 (1979).
80. Y. Einaga and H. Fujita, *Reorji Gakkaishi* **12**, 136 (1984).
81. W. Hess, *Macromolecules* **20**, 2587 (1987).
82. K. L. Ngai, A. K. Rajagopal, and S. Teitler, *J. Chem. Phys.* **88**, 5086 (1988).
83. C. A. Angel and M. Goldstein (Eds.), *Dynamic Aspects of Structural Change in Liquids and Glasses*, New York Academy of Sciences, New York, 1987, Vol. 284.
84. G. B. McKenna, K. L. Ngai, and D. J. Plazek, *Polymer* **26**, 1651 (1985).
85. K. L. Ngai, A. K. Rajagopal, and T. P. Lodge, *J. Poly. Sci., Poly. Phys. Ed.* submitted.
86. K. L. Ngai and T. P. Lodge, *J. Chem. Phys.* submitted.
87. R. S. Adler and K. Freed, *J. Chem. Phys.* **72**, 4186 (1980).
88. M. Daoud, J. P. Cotton, B. Farnoux, G. Jannink, G. Sarma, H. Benoit, R. Duplessix, C. Picot, and P. G. de Gennes, *Macromolecules* **8**, 805 (1975).
89. J. G. Kirkwood and J. Riseman, *J. Chem. Phys.* **16**, 565 (1948).
90. L. M. Wheeler and T. P. Lodge, *Macromolecules* **22**, 3399 (1989).
91. T. P. Lodge, P. Markland, and L. M. Wheeler, *Macromolecules* **22**, 3409 (1989).

AUTHOR INDEX

Numbers in parentheses are reference numbers and indicate that the author's work is referred to although his name is not mentioned in the text. Numbers in *italic* show the pages on which the complete references are listed.

SUBJECT INDEX